Manager's Guide to Preventive Building Maintenance

Manager's Guide to Preventive Building Maintenance

Ryan Cruzan

LONDON AND NEW YORK

Published 2020 by River Publishers
River Publishers
Alsbjergvej 10, 9260 Gistrup, Denmark
www.riverpublishers.com

Distributed exclusively by Routledge
4 Park Square, Milton Park, Abingdon, Oxon OX14 4RN
605 Third Avenue, New York, NY 10017, USA

First issued in paperback 2023

Library of Congress Cataloging-in-Publication Data

Cruzan, Ryan, 1970-
Manager's guide to preventive building maintenance/Ryan Cruzan.
p. cm.
Includes index.
ISBN-10: 0-88173-619-8 (alk. paper) -- ISBN-13: 978-8-7702-2281-5 (electronic)
ISBN-13: 978-1-4398-1431-4 (Taylor & Francis: alk. paper)
1. Buildings--Maintenance. 2. Building management. I. Title.

TH3351.C78 2009
658.2'02--dc22

2009008439

Manager's guide to preventive building maintenance/Ryan Cruzan.
First published by Fairmont Press in 2009.

Routledge is an imprint of the Taylor & Francis Group, an informa business

Publisher's Note
The publisher has gone to great lengths to ensure the quality of this reprint but points out that some imperfections in the original copies may be apparent.

ISBN 13: 978-87-7022-904-3 (pbk)
13: 978-1-4398-1431-4 (hbk)
13: 978-8-7702-2281-5 (online)
13: 978-1-0031-5150-0 (ebook master)

Dedication

To my children, Sarah and Scott: Thank you for putting up with the many evenings and weekends I spent working on my "stupid book" again.

...and to my wife, Kim, who also tolerated (or maybe didn't really mind so much) my being busy with this for so long. Thank you for taking the time to edit this over and over again when I know you really would have preferred reading something with more suspense and romance.

Table of Contents

PART I—Creating an Effective PM Program 1

Chapter 1—What is PM? 3
So, What is PM? 4
Preventive maintenance defined 4
Why Do PM? 5
Reason 1: PM extends equipment life 5
Reason 2: PM reduces costs 5
Reason 3: PM saves energy 7
Reason 4: PM improves the experience of your occupants. 7
Reason 5: PM makes your job easier. 8
Predictive Maintenance (PdM) 9
Thermal imaging 10
Group re-lamping 10
Infant mortality and the bathtub chart 11
Other types of maintenance 13
Corrective maintenance 13
Deferred maintenance 15
Run to failure 15
Emergency maintenance 16
Call-back maintenance 16
PM planning begins with design 17

Chapter 2—The Economics of Preventive Maintenance 21
The Jones-Lang-LaSalle Report 22
Selling PM to Management 23
PM as a Budgeting Tool 27
The Long Range Facilities Plan 27
The Repair or Replacement Decision 28
The Useful Life of Building Systems 33
Estimating PM Costs 34

Chapter 3—Getting to Work—Setting Up a PM Program 37
Too Busy for Preventive Maintenance 37
Not Enough Maintenance Staff 37

Can't Afford to Hire Skilled Staff for Specialized Tasks 38
Don't Need to PM Something That's Already Past Its
Useful Service Life 38
Starting To Do Preventive Maintenance 38
Step 1—Determining Your Organization's Goals 39
Step 2—Inventory Capital Equipment and Decide What to PM 40
Equipment Data Sheets 42
Unique Identifier 44
Commonly used equipment identifiers 45
Step 3—Scope and Frequency 48
Deciding what PM tasks to do and how often to do them 48
Scope—or how will you PM each piece of equipment? 48
CLAIR 50
Frequency—or how often will
you do PM to each piece of equipment? 51
PM schedules based on hours of operation 52
Choosing a CMMS 53
Do I really need a CMMS? 54
Step 4—Making a PM Calendar 57
Scheduling by the week vs. by the day 59
Other things on my PM Calendar 60
Getting your Hands Dirty—
The first few month of actually doing PM 60
Getting Support from Upper Management 61
Document, Document, Document 63
As a reference. 63
As a CYA. 63
As a CYCoA. 63
Regulatory compliance 64

Chapter 4—The people that Do PM 67
Managing People 68
Getting Staff to Buy into Your Program 69
That Attitude of Continual Improvement 70
Training and Direction 70
Where to get staff trained 71
An HVAC example 72
Investing in Tools 73
Trust but Verify 74

Outsourcing PM 75
Final Thoughts 76

PART II—Technical Information for Preventive Maintenance Success 79

Chapter 5—Lubrication Theory 83
Tribology 83
Reynold's Theory 83
Reynold's Theory Equation 86
Choosing a Lubricant 89
Viscosity 89
Problems with choosing the wrong viscosity. 90
Types of Lubricants 91
Mineral oils 91
Synthetic oils 92
Solid lubricants 92
Greases 92
The Useful Life of Oil and Grease 94
How Often Do We Need to Grease Bearings? 95
Types of Bearings 98
Plain or sleeve bearings 98
Rolling element bearings 99

Chapter 6—Maintaining Commercial Roofs 103
The Four Common Types of Commercial Roofs 104
Built-up roofing (BUR) 105
Modified bitumen roofing (MBR) 106
EPDM and thermoplastics 107
Standing seam metal roofing 109
Roof Insulation 110
Common types of roof insulation 112
Fasteners, Flashing, and Roof Penetrations 113
Fasteners 114
Flashing 114
Roof Warranties 116
Preventive Maintenance of the Roof 118
Built-up Roofing (BUR) & Modified Bitumen Roofing (MBR) and Single Ply (EPDM) PM Inspection Checklist 120

Flashing, fasteners, penetrations PM inspection checklist 121
Standing seam metal roofing PM inspection checklist 121
All roofing types PM inspection checklist 121

Chapter 7—HVAC Systems 125
Refrigeration Machinery 126
Heat Pumps 129
Preventive Maintenance of Air Conditioning and
Refrigeration Equipment 130
Preventive maintenance of condensing units—compressors 131
Preventive maintenance of condensing units—
The condenser coil 133
Preventive maintenance of air handlers—the evaporator coil......... 134
Preventive maintenance of air handlers—air filters 135
Preventive maintenance of air handlers—
condensate equipment 136
Preventive maintenance of air handlers—blowers......................... 137
Preventive maintenance of air handlers—
heating and other equipment 138
Cooling Towers and Cooling Loops 139
Preventive maintenance of cooling towers and cooling loops......... 142
Chillers 142
Automated Controls 144
HVAC PM Checklist 146

Chapter 8—Belt Drives 149
The 3 Types of Drive Belts.. 150
Drive Belt Lengths.. 151
What If I Don't Have an Old Belt to Measure? 152
Other Drive Belt Characteristics.. 153
Proper Operation.. 154
Problems with Belt Drives 155
Inspecting Belt Drives... 156

Chapter 9—Indoor Air Quality 159
The 3 Parts of IAQ Problems (The 3 P's)... 159
Pollutants—the first P 160
The most common IAQ pollutants... 160
Pathways—the second P 161

People—the third P 162
Comfort Issues 163
Solving IAQ Problems 164
Pollutants—solving problems 165
Pathways—solving problems 165
People—solving problems 166
The 8 common pollutant solutions 167
The "M" Word 167
Difficult IAQ Problems—Sick Building Syndrome 168
Collecting Data 169
Seeking Professional Assistance 170
Effective Communication 172
PM Tasks That Affect IAQ 172

Chapter 10—Paint and Protective Coatings 175
Types of Paint 175
Oil-based paint 177
Latex paint 178
Elastomeric wall coatings 179
Epoxy paints 180
Urethane paints 180
Rust inhibitive paints 181
Mildew resistant paints 181
Cold galvanizing compounds 181
Primers and Sealers 182
Water and oil-based primers 183
Shellac primers 183
Rust-inhibitive primers 183
Bonding primers 184
The Right Paint or Primer for the Job 184
Preparing the Surface 185
Note about lead paint: 186
Six steps to proper surface preparation 186
Rust converters 187
Environmental Concerns 188
Volatile organic compounds (VOCs) 188
Lead paint 189
Paint disposal 189
Paint Failures 191

Application temperature 191
Common modes of paint failure 192
Inspecting and Recoating 194

Part III—Specific Maintenance Procedures and Requirementsı 197

Chapter 11—Specific Maintenance Procedures and Requirements 199
List 1: PM Tasks Listed by Frequency 201
List 2: PM Tasks required by building codes or regulatory agencies... 204
List 3: Equipment Specific Procedures, Requirements,
and Technical Details for PM 206
Informational Symbols 206

APPENDIX 283
Sample Preventive Maintenance Record Forms 283
Emergency generator testing log 284
Hotel room PM checklist 285
Smoke detector testing log 287
Troubleshooter's Creed 288
Truisms 289
Glossary of Preventive Maintenance Terminology 290

INDEX 299

Preface

Manager's Guide to Preventive Building Maintenance is written for the building manager wanting to improve the condition of their properties while reducing costs and making the work of the facilities department more manageable. This book will provide the tools necessary to create and implement an effective preventive maintenance program in your facility.

Preventive maintenance (PM) has been a familiar principal within industry and manufacturing for several decades. Comprehensive preventive maintenance programs have only recently become a common part of the maintenance program within facilities. Today, a large percentage of facilities still have no PM program in place or have PM programs which include little more than semi-regular changing of air conditioning filters. Each year, more and more well managed facilities recognize the benefits of a comprehensive program of PM and understand how significantly a well designed PM program can affect every aspect of a company's operations.

A well designed and well implemented PM program will extend the life of equipment by reducing replacement costs, preventing most breakdowns that effect building occupants, maintaining the energy efficiency of equipment, improving the effectiveness of the maintenance department, and improving the overall condition of a building and the experience of its occupants.

My first experience with preventive maintenance was nearly 20 years ago working for a hotel that did not have a preventive maintenance program. I spent long hours and late nights running from emergency to emergency. After reading an article in a trade publication about the benefits of PM, I created a simple PM program and was amazed how this changed my daily work routine. After only a couple of months of diligently following my simple schedules, emergencies almost stopped, guest complaints were suddenly gone, and I was working with less stress and getting more sleep. Since that early experience, I have created and refined many successful PM programs at facilities in several industries.

Most of the few books available on the subject of the preventive maintenance of buildings are written by professionals who have extensive experience with PM as viewed from the office or spreadsheet. I wrote this book because I believe I offer the unique perspective of someone who has actually spent a career doing and supervising the daily activities of the maintenance department.

Two years ago, as a requirement toward earning certification as an Educational Facilities Manager, I took a course on preventive maintenance through the state university. The textbook was disappointingly basic and not especially helpful in the real world of facilities maintenance. I had considered someday writing a book about PM and it was this class which convinced me I had something to offer and swayed me to put pen to paper.

It is my hope that this book will serve the building manager in creating an effective preventive maintenance program that works well in this challenging field of facilities management. I hope the facilities manager who entered the field from academia will gain perspective into the nuts and bolts world of the maintenance department where PM is performed. I also hope those managers who rose through the ranks from the maintenance department will gain a stronger understanding of the engineering fundamentals behind the PM tasks they perform.

But mostly I hope that, by reading this book, your buildings will last longer, your work load will become more manageable, emergencies will become fewer, maintenance costs will decrease, department morale will increase, and you will enjoy your work more and get more sleep at night.

Introduction

Managers Guide to Preventive Building Maintenance has been written in three parts. The first part discusses components of an effective PM program and gives the reader the necessary tools to create one in their own facility. The second part explains the science and engineering principals behind many of the most common and important PM tasks such as lubrication theory, HVAC system maintenance, how PM impacts indoor air quality, and the selection and application of paints and architectural coatings. The third section is a comprehensive reference section including over 100 different types of equipment commonly found in facilities including theories of operation, maintenance requirements, and government codes or regulations that apply to the maintenance of each equipment type.

Part 1—Creating an Effective PM Program

The first section of this book (Chapters 1 through 4) defines preventive maintenance and explains why and how to set up a PM program. Some of the many advantages preventive maintenance brings to facilities are discussed. Other types of maintenance such as corrective maintenance, deferred maintenance, and emergencies are all reduced when an effective PM program is being followed.

Also included in part 1 is the return on investment (ROI) of a good PM program and why investing in PM is often one of the most lucrative investments a company can make. A dollar saved through PM is as good as a dollar earned through any other business activity and few business activities can come close to the ROI possible through PM. The reader will also learn how PM can be used as a budgeting tool, how to accurately make repair verses replace decisions for equipment and how to estimate PM costs.

Developing a successful PM system from goal setting and equipment inventory through determining the exact procedures to follow for each piece of equipment is explained. Setting up an intelligent PM schedule or calendar, and gaining the support and necessary start up financing from upper management is also included.

This section also addresses the selection and management of the people who will be performing the actual hands on work of PM. Included are managing staff and getting staff to support a new PM program, effective training options, and outsourcing all or part of your PM program.

Part 2—Technical Information for Preventive Maintenance Success

The second part of *Manager's Guide to the Preventive Maintenance of Buildings* is written to help the facilities manager with a technician background to become familiar with the most important science and engineering principles governing the PM tasks we do. This section is also intended to serve as a training reference for line staff performing hands on PM work.

The six chapters included in part two of this book address the science of lubrication theory, commercial roofing maintenance procedures, maintaining HVAC systems, belt drives, indoor air quality, and a chapter on protecting buildings using paint and architectural coatings. These six technical areas offer the most impact for a PM program and are trade areas loaded with misinformation and inaccurate on-the-job folklore. The purpose of part two is to dispel the myths and provide useful information to make your PM program more effective.

Part 3—Specific Maintenance Procedures and Requirements

The final section of this book is found in Chapter 11. This chapter serves as a desk reference of preventive maintenance tasks and procedures for nearly a hundred different types of equipment. Maintenance tasks, engineering principles, operating principles, and the details of applicable codes or regulations are included for each piece of equipment.

Part I

Creating an Effective PM Program

PREVENTIVE MAINTENANCE DEFINED

> **"Preventive maintenance is a scheduled program of regular inspections, adjustments, lubrication, or replacement of worn or failing parts in order to maintain an asset's function, and efficiency."**

Part one of *Manager's Guide to Preventive Building Maintenance* discusses why and how to set up an effective preventive maintenance program for any facility.

Chapter 1: What is PM

PM includes all the tasks we perform to keep our buildings in good condition. Oiling or greasing bearings, changing filters, painting, and other tasks are the nuts and bolts work of a good PM program. The incidents of breakdowns and the need for other types of maintenance such as corrective maintenance or emergency maintenance is reduced when a successful PM program is in place.

Preventive maintenance extends equipment live, reduces breakdowns, saves money, improves the experience of building user, and makes the work of the maintenance department more manageable. When the maintenance department starts a program of preventive maintenance, its focus shifts from one of fire fighting to one of planned and predictable maintenance tasks.

Chapter 2: The Economics of PM

Business decisions are made according to the anticipated return on investment (ROI). preventive maintenance offers average returns of 500% or more and is one of the most lucrative investment opportunities a business can undertake. Specific maintenance tasks can have ROIs of over 2000%. A dollar saved through PM is as good as a dollar earned from any other business activity.

Chapter 3: Setting Up a PM System

Determining an organizations goals, making a thorough equipment inventory, setting up a PM schedule or calendar, and developing a system of record keeping are the basic parts of creating a PM program. To create a good PM program, it is necessary to know the specific maintenance requirements for each type of equipment included in the program. Some organizations work well with a paper system of schedules and record keeping. Larger organizations may prefer to keep schedules and maintenance records in a computerized maintenance management system (CMMS).

Chapter 4: The People that do PM

The most important aspect of your PM program is your maintenance staff. Your maintenance staff's skills, training, and most importantly, attitude, can make or break a good PM program. To be successful, your staff will need training, tools, and knowledge. Starting a PM program is difficult because it means additional work that needs to be completed until the program is given a chance to work and breakdowns and emergency repairs decrease. Eventually, a PM program will make the work of the maintenance department easier and more manageable. In the short term, there will be some growing pains.

Chapter 1

What is PM?

As facilities maintenance professionals, we will never have to worry about our industry becoming obsolete. All building components will decay, wear away, or otherwise fail eventually. As long as buildings exist: steel will rust, glass will break, motor bearings will wear, filters will get dirty, ballasts will burn out, pipes will corrode, and roofs will leak. As facilities maintenance professionals it is our job to keep these building components working for as long as possible and to make sure the inevitable equipment failures are rare occurrences. We are charged with keeping the normal effects of deterioration and wear to a minimum. The way to keep deterioration at bay is through an effective program of preventive maintenance.

The benefits of a good preventive maintenance (PM) program are huge compared to the work involved in setting up a good program. Still, few facilities have good PM programs. I have worked for facilities departments in several industries as a maintenance mechanic or maintenance supervisor. I have worked in buildings with no PM programs, in buildings with poorly designed PM programs, and in buildings with good PM programs. I can assure you that having a good PM program is worth every bit of work involved in getting the program in place.

It has also been my experience that there are few people working in facilities that know very much about setting up a good PM program. Most facilities people know PM is important, but do not really know where to start. This book is going to show you how to create a simple yet effective PM program, tailored to your facility that will work to save maintenance costs, improve equipment performance, improve the experience of building occupants, extend equipment life, and most importantly: make your job easier.

SO, WHAT IS PM?

Preventive maintenance, is the normal, everyday work we do to protect the condition of our properties and to prevent equipment failure that normally occur within a facility. PM includes all of the tasks we perform to keep a building and its equipment in good condition. Preventive maintenance includes changing heater filters, checking drive belts for wear, checking oil levels, inspecting roof flashing, greasing bearings and painting window trim. PM extends equipment life, keeps equipment running efficiently, and reduces breakdowns. Doing simple, inexpensive PM today saves time and money that would otherwise be spent doing major repairs or replacing equipment tomorrow.

Preventive Maintenance Defined

> **"Preventive maintenance is a scheduled program of regular inspections, adjustments, lubrication, or replacement of worn or failing parts in order to maintain an asset's function, and efficiency."**

Preventive maintenance is intended to keep minor problems from escalating into major problems. Preventive maintenance allows a maintenance department to transition from a fire-fighting approach of running from one emergency breakdown to another to preventing those breakdowns before they occur.

Hopefully, you noticed the above definition includes the words "scheduled" and "regular." An effective PM program will ensure that everything in your building is seen on a regular schedule; whether it seems to need it or not.

A preventive maintenance schedule is basically a calendar of PM tasks to be performed. These tasks will be things such as quarterly air conditioning filter changes, monthly roof inspections, weekly checks of lawn irrigation systems, and seal coating the parking lot annually. Preventive maintenance tasks are done at regularly scheduled intervals to prevent future equipment problems.

If we wait for some piece of equipment to tell us it needs attention, that's a repair, not PM. Repairs are expensive and time consuming. preventive maintenance tasks are generally low-cost and don't take much time. Many times, expensive, time consuming repairs need to be done because we failed to do the cheaper and faster preventive maintenance.

I am sure that we have all had the experience of repairing some type of equipment, for example a circulating pump, and found failed bearings that were completely dry and empty of lubricating grease. And you probably thought that if someone had just greased these bearings once-in-a-while, you wouldn't be making this repair now. That's what happens when preventive maintenance is neglected. Corrective maintenance becomes necessary.

The above definition of preventive maintenance says that we do PM on a building's "assets." An asset can be any building equipment or component. Assets include air conditioners, emergency generators, roofs, steam boilers, and rain gutters. Assets can be stationary, fixed, permanent parts of the building or items that are not a fixed part of a building. Items such as lawnmowers, hospital beds, or laboratory equipment are also assets. Nearly every type of equipment can benefit from some type of preventive maintenance.

WHY DO PM?

We have already briefly touched on a few reasons to do PM. Here are some others:

Reason 1: PM Extends Equipment Life

Probably the biggest reason to do preventive maintenance is that it keeps equipment running longer. The most familiar PM activity for most people is having their automobile serviced. Checking the oil level and changing the oil in our car is a PM task we all do fairly regularly. Hopefully, we do not wait until the family car breaks down to change the oil. We change the oil every 3,000 miles, or as often as the manufacturer recommends, to keep the car's engine in good condition. We do this even when the car is running well.

We know that if we don't follow the manufacturer's recommendations and service our car regularly we should expect a catastrophic engine failure within the first couple of years. In contrast, by just checking and regularly changing the oil, we can easily extend a vehicles service life to a decade or more. None of us would wait until the engine seized from lack of lubrication before we changed the oil but lots of maintenance departments operate that way.

Reason 2: PM Reduces Costs

We know that we can extend the service life of equipment through preventive maintenance. It should be obvious that extending the service life of equipment saves money. When equipment lasts longer, you do not have to buy replacement equipment as often. This reduces the long-term cost of owning the equipment. It does not take many years to realize the savings of maintaining a cooling tower and replacing it after 15 years verses ignoring it and replacing it after five years.

Preventive maintenance can also reduce costs by reducing the expense of hiring outside contractors. In all but the largest organizations, maintenance departments occasionally rely on outside contractors with specialized skills. If PM is not being performed consistently by your in-house staff, problems that often could have been prevented in-house can become expensive problems requiring the more costly repair services of these outside contractors.

One of the most familiar examples of this happens very often with air conditioning. If an air conditioner's air filter is not changed regularly, the filter will eventually become clogged with dust, blocking air flow through the evaporator coil. When this happens, the evaporator coil can freeze due to the reduced air flow. An air conditioner with a frozen coil will no longer provide cooling and condensate will drip from the ice creating a puddle on the floor. If a small maintenance department does not have staff certified by the EPA to work on air conditioners, an outside contractor will need to be called to make repairs.

With travel time charges, and hourly minimum charges, it can easily cost $200 or more to have a technician change a filter that you could have changed in-house for less than $10.00 including labor and materials. I've seen "frozen coil—changed filter" written on service tickets more times than I would like to admit.

Equipment downtime, or the time that equipment is not working, can also be a source of increased costs. Proper preventive maintenance of equipment will reduce downtime. I once had the embarrassment of shutting down 40 occupied hotel rooms because an air conditioning unit failed. The compressor had a slow oil leak that we had not seen which eventually caused the compressor to fail. The hotel I worked for catered to a clientele of corporate executives who were usually in our area on extended business.

Because we overlooked this oil leak, the hotel lost the night's room revenue on 40 rooms; had to hire a shuttle bus to move everyone to a competing hotel for the night; had to pay for 40 guest rooms at the competing hotel; had to hire the shuttle bus again the following morning to get everyone back to our hotel so everyone could get ready for work; and offered everyone some very nice and expensive complimentary meals, drinks, and outings for their trouble. Half of our inconvenienced guests decided to stay at the competing hotel and we were left with empty rooms for several nights. Since hotels make their money on "heads in beds," these empty beds were revenue losers.

We had been changing AC filters regularly but were not doing routine inspections to look inside the cabinets to check for any obvious problems. If we had been inspecting these properly, we would have noticed the puddle of oil weeks before the compressor failed. We saved the 10 minutes of work that it would have taken to do the inspection. We lost thousands of dollars of room revenue as a result. You can bet I do visual inspections on AC units as part of my PM program now.

Reason 3: PM Saves Energy

Energy costs can also be reduced by simple PM tasks. With the recent increase in the cost of natural gas, electricity, and fuel oil, the energy savings created by preventive maintenance are more important than ever. Slipping drive belts, dirty electric motors, and clogged air filters all cause increased energy usage and are easily correctable through proper PM.

In the previous example of a dirty air conditioning filter, not only would such an ac unit freeze up but the efficiency of the unit would decrease dramatically during the last few months the unit struggled to perform with such dramatically reduced air flow. A modern high-efficiency air conditioner with a dirty filter will no longer perform at its designed high-efficiency. Replacing a $5.00 filter and cleaning a clogged evaporator coil can reduce the amount of electricity used by 50% or more. Regular filter changes can be expected to reduce overall air conditioning operating costs by 8 to 10%. For large facilities, this can mean thousands of dollars in energy savings each season.

Other examples of PM tasks that can save energy are: inspecting roofs for wet insulation which allow thermal losses, maintaining window caulk, inspecting weather-stripping on doors, and making sure that automated building comfort systems are operating properly.

Reason 4: PM Improves the Experience of Your Occupants.

Whether you maintain a retail space, a healthcare facility, a hotel, a schools, an office building, or some other type of facility; you have control of small details that can either make your building occupant's experience positive or negative. Air conditioning or heating systems that fail regularly, parking lots littered with broken glass, roofs that leak, fire alarm systems with frequent false alarms, and rest room partition doors that do not latch are just a few examples of the small but very frustrating experiences that can be solved with a simple PM program of scheduled inspections and repairs.

If your tenants are regularly complaining about maintenance problems that need attention, this is a sign that you need a PM program. With a good preventive maintenance program in place and working for you, the maintenance department can become almost invisible since you will be working behind the scenes to keep things running without breakdowns instead of working to fix tenant complaints after breakdowns occur. Once a PM program is in place, you will find and solve these problems before your building occupants do.

Reason 5: PM Makes Your Job Easier.

We know that PM extends equipment life, saves your company money, saves energy, and improves the experience of your building's occupants. As the maintenance manager you are probably wondering "What's in this for me?"

Here is what you and your department can get out of a good PM program:

- Fewer midnight emergency phone calls
- Fewer weekends and late nights at work
- Fewer angry phone calls from dissatisfied building occupants
- Less stress
- More satisfaction and pride in the improved condition of your facility.

I have set up PM programs at several properties during my career. I can tell you from personal experience that starting a PM system will be a lot of work in the beginning but the extra work is definitely worth it in the long run. In the beginning it will be hard to find the time to do all of the preventive maintenance tasks on your new schedule. Lots of maintenance departments realize the importance of PM but don't do any, because they are just too busy running from emergency to emergency to find the time. Keep your eye on the prize because if you can stick to your PM schedule for one complete cycle, typically 3 months, you will see a sudden and dramatic decrease in these emergencies. It works and it is worth every bit of extra work required in the beginning.

Even with all the benefits a PM program provides to the bottom line, building occupants, and the maintenance department, it is estimated that only 15% of buildings have comprehensive preventive maintenance programs.

PREDICTIVE MAINTENANCE (PDM)

PdM stands for predictive maintenance. Predictive maintenance is similar to PM (preventive maintenance) in many respects. You will sometimes hear the terms PM and PdM interchanged. Like preventive maintenance, predictive maintenance is used to keep equipment in good repair and fix problems before the equipment fails. The difference between the two is that PM is time-based while PdM is condition-based. That means that PM tasks are scheduled according to a calendar while PdM tasks are scheduled when indicated by some sort of measurable wear factor.

Changing air filters every 3 months would be considered PM since the schedule is time-based—every 3 months. Changing air filters only when the filters are getting dirty would be PdM. Many commercial air conditioners offer this feature. Many AC units have pressure sensors in the air filters compartment and will flash the word "filter" on the thermostat display when the air filters are beginning to get dirty and the air flow is starting to become obstructed. Because the filter change interval depends on a measured reduction in air flow, changing these filters would be considered PdM instead of PM.

PdM requires constant monitoring of equipment conditions. PdM offers some cost savings because maintenance tasks are only performed when needed. However, the costs of continuously monitoring the condition of equipment often outweigh the savings.

PdM is more common in industry and manufacturing maintenance than it is in facilities maintenance. An entire field of reliability engineering has developed in the manufacturing sector which uses techniques such as vibration analysis, thermal imaging, oil analysis, and ultrasonic detection to inspect all sorts of equipment to predict component failures before they occur. All of these expensive techniques make sense in a manufacturing environment where a single machine can cost millions of dollars and where machine down time can cost thousands of dollars in lost production every hour.

Facilities maintenance does not involve the same types of costly one-of-a-kind machinery found in industry. Facilities equipment is also less complex and less prone to failure than the custom-built machinery found in manufacturing plants. This is why PdM is more often used in industry and manufacturing than it is in facilities. However, there are some predictive maintenance technologies that have been borrowed from industry that fit well with preventive maintenance of buildings.

Thermal Imaging

Thermal imaging is becoming more commonplace in facilities maintenance. A thermal imaging camera is able to record temperature in the same way a regular camera records color. Thermal cameras can identify problems by letting the camera operator actually see hot and cold.

The primary use for thermal imaging in facilities maintenance is to detect electrical problems before they cause failures. Corroded or loose electrical connections cause a point of high electrical resistance. High resistance points will overheat as electrical current passes through them. Overheating can eventually lead to melted wire insulation, damaged or tripping circuit breakers, blown fuses, and other heat related damage to electrical equipment. A thermal imaging camera can see these hot spots long before damage is done.

Thermal imaging is also used to detect defects in the building envelope. The building envelope encompasses roofs, walls, windows, or doors. Thermal imaging can display the heat or cooling energy losses and tell you exactly where you have cracks or missing insulation. Warm or cool spots on your building indicate that your heating or cooling energy is escaping at these locations.

Contractors performing roofing inspections often use thermal imaging cameras to detect water under the roof's surface. During the heating of day, or cooling of night, water saturated insulation trapped under the roof surface will maintain its temperature longer than dry insulation. By looking at thermal images of the roof, it is possible to detect hidden areas of water damage without damaging the roof by taking core samples. In the same way, thermal imaging can be used to find water in concrete block, brick, ceilings, or carpets. Early detection of water can be valuable in keeping the building structure from deteriorating.

Group Re-lamping

With fluorescent or high intensity discharge (HID) lighting, it can make sense to re-lamp an entire facility at one time rather than changing bulbs as they burn out. Fluorescent and HID bulbs tend to have similar expected lifetimes so it can be assumed that the majority of these bulbs will fail at nearly the same age. If we know at what age the lamps can be expected to fail, we can change a group of lights together just before this mass failure occurs. Fortunately, lamp manufacturers are able to provide expected lifetimes for their lamps under a variety of operating conditions.

Another reason to re-lamp a facility is a trait of all HID lamps known as lumen maintenance. As HID lamps age, their light intensity declines. Some HID lamps can loose 40% of their light output by the time they reach the end of their service life. In many cases, a facility may decide to re-lamp to maintain the original light levels. Figure 1-1 shows the expected life and lumen maintenance of several common types of lamps.

Group re-lamping is considered PdM since a measured number of bulb failures (by the manufacturer in field testing) or a measured decrease in light intensity is used to determine when to schedule the activity.

Facilities that group re-lamp tend to do so roughly every five years and are able to eliminate all of the man hours spent setting up ladders or other equipment to change bulbs every day as they burn out. Areas such as conference centers, warehouses, auditoriums, gymnasiums, and highway signs often have difficult access to lighting due to their height. Setting up cranes or aerial lifts once every five years instead of every time a bulb blows out makes group re-lamping an attractive option.

As you can see, some of our planned maintenance activities are actually PdM rather than PM. However, the two terms are very closely related and the term preventive maintenance is more common in the field of facilities maintenance. We will use the term PM whether an activity is condition-based or time-based to describe all planned maintenance activities.

Infant Mortality and the Bathtub Chart

If you are considering group re-lamping for your facility, you should be familiar with a phenomenon that reliability engineers call "infant mortality." The concept is simple: Due to manufacturing defects or installation errors, new equipment is more likely to fail than equipment that has already survived its "burn in" period. This premise applies to all new equipment, including lamps. Engineers call a graph of failure rates over time, as shown in Figure 1-2, a "bathtub chart" because of the shape of the curve.

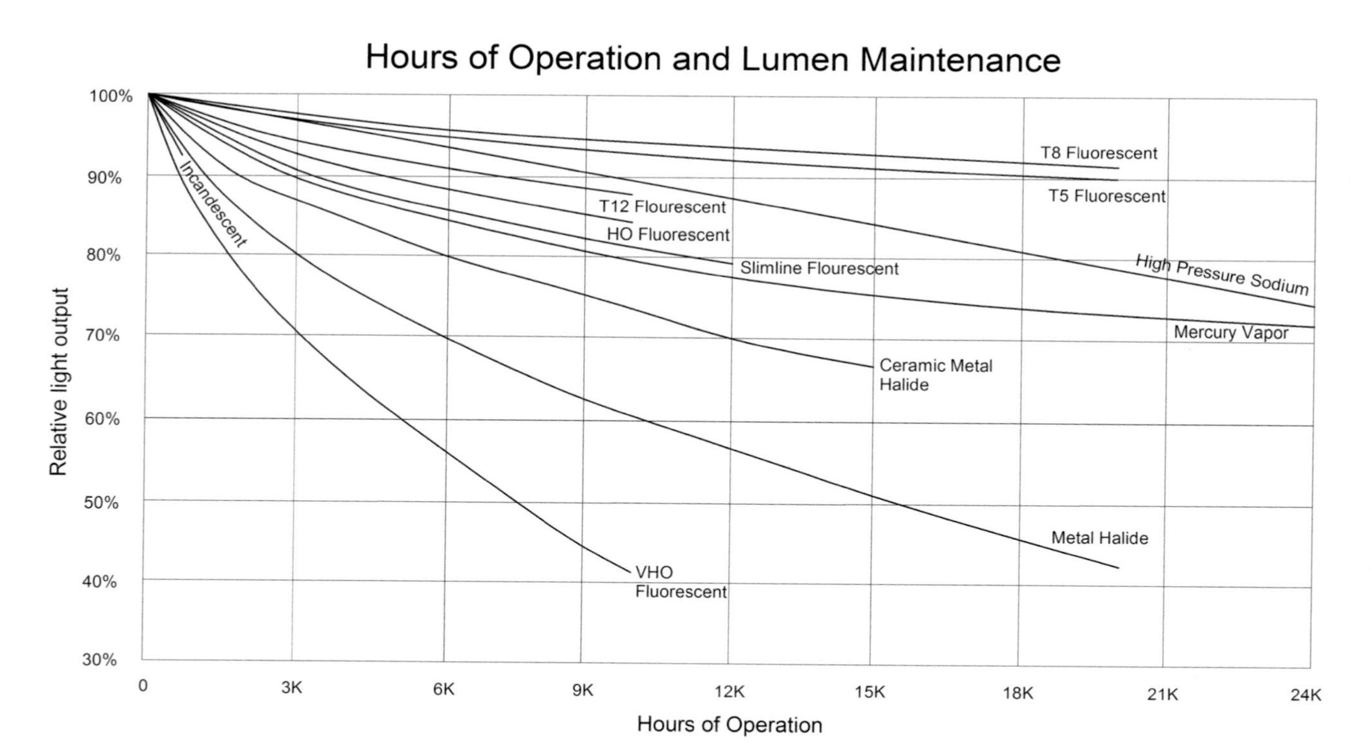

Hours of Operation and Lumen Maintenance
Relative light output
100%
90%
80%
70%
60%
50%
40%
30%
0
3K
6K
9K
12K
15K
18K
21K
24K
Hours of Operation
T8 Fluorescent
T5 Fluorescent
T12 Flourescent
HO Fluorescent
Slimline Flourescent
High Pressure Sodium
Mercury Vapor
Ceramic Metal Halide
Metal Halide
VHO Fluorescent
Incandescent

Figure 1-1

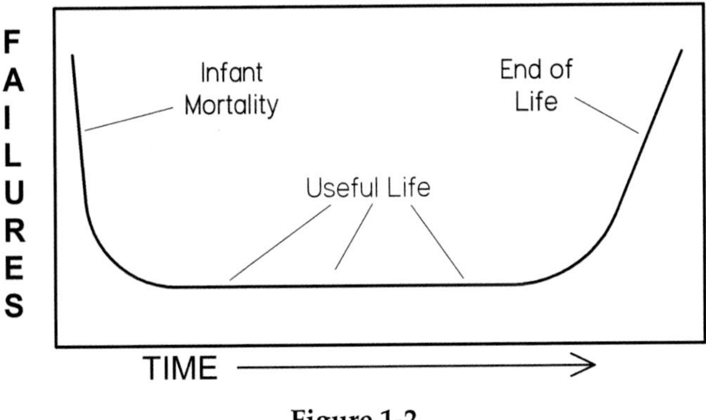

Figure 1-2

If all the lamps in a facility are replaced at the same time, there will be a brief period of a few weeks when there will be a high number of lamp failures (left side of chart) This is because any factory defects will show up when the lamps are first put into use. After this initial "burn in" period, when all the defective lamps are weeded out, the failure rate will drop considerably and will remain relatively constant for several years (middle of chart) with a few failures expected each year.
Near the end of the lamps' expected life, the failure rate will increase again as lamps fail due to age (right side of chart)

OTHER TYPES OF MAINTENANCE

There are many types of maintenance tasks performed by the facilities maintenance department. In fact, many organizations are missing out on the benefits of PM and do very little PM at all. Figure 1-3 shows a flowchart of the different types of maintenance done by maintenance departments. These types are discussed in detail below.

Corrective Maintenance

Corrective maintenance is also known as "reactive maintenance" or just "repairs." Corrective maintenance is fixing something that is already broken. This type of maintenance is probably the most common type of maintenance done in most facilities. In organizations that do not do PM or do very little PM, the amount of corrective maintenance can become overwhelming. The purpose of a PM program is to reduce the amount

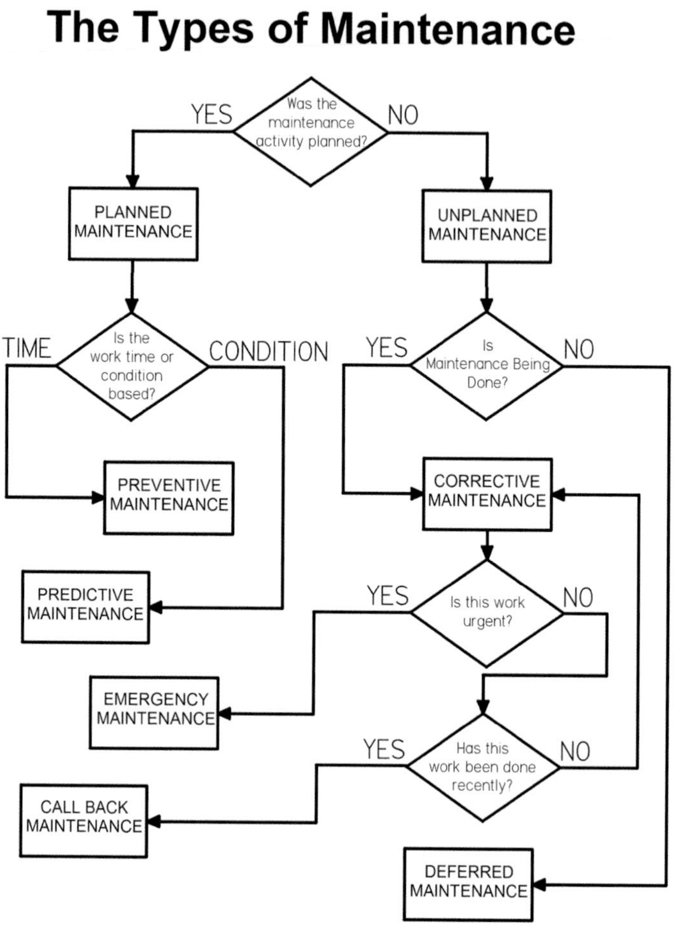

The Types of Maintenance

Was the maintenance activity planned?
YES
NO
PLANNED MAINTENANCE
UNPLANNED MAINTENANCE
Is the work time or condition based?
TIME
CONDITION
PREVENTIVE MAINTENANCE
PREDICTIVE MAINTENANCE
Is Maintenance Being Done?
YES
NO
CORRECTIVE MAINTENANCE
Is this work urgent?
YES
NO
EMERGENCY MAINTENANCE
Has this work been done recently?
YES
NO
CALL BACK MAINTENANCE
DEFERRED MAINTENANCE

Figure 1-3

of corrective maintenance that needs to be done. Even with an effective PM program is in place, corrective maintenance can never be completely eliminated.

As its name implies, corrective maintenance is any maintenance that corrects a problem; as opposed to PM that attempts to prevent problems before corrective maintenance is necessary. A few familiar corrective maintenance tasks would include patching leaking roofs, replacing motors with seized bearings, and replacing burned out fluorescent bulbs.

Most corrective maintenance comes to the maintenance department from complaints by building users. These complaints may be phone calls requesting that repairs be made or can be in the form of paper or computerized work orders. Corrective maintenance needs to be performed timely to keep building users content but cannot get in the way of performing PM tasks. One of the hardest parts of managing a maintenance department is to stay on-task with scheduled maintenance while other emergencies are coming in. I am not suggesting that you ignore the corrective maintenance that needs to be done, only that you keep PM as your priority. In the long term, PM will have a much larger impact on building condition (and therefore building occupants) than corrective maintenance.

Deferred Maintenance

Deferred maintenance is another way of saying "no maintenance." Deferred implies that we will do some maintenance in the future but not today. Maintenance is often "deferred" when there are budget crisis.

If it is expected that maintenance funding will be available at some date in the future, the costs of needed repairs should be recorded as breakdowns occur. Knowing an accurate cost of "catching up" will be a valuable budgeting tool in the future.

If an organization is anticipating moving to a new location or closing a local branch, maintenance may be deferred so that money is not spent on maintaining assets that will soon not be needed. In this case, the term "deferred" is still used although there are no real intentions of doing the needed work later.

Run to Failure

A "run to failure" policy may make sense for very inexpensive equipment or building components that can be replaced more cheaply than they can be repaired. Small office tools such as pencil sharpeners or coffee makers would most likely not be part of your PM program. If you were to disassemble and lubricate and vacuum out the tiny motor on a pencil sharpener, you could likely extend its service life. But you can replace the entire unit for less than the cost of labor to do the PM. Good judgment would be to choose "run to failure" as the best course of action. A desktop pencil sharpener might be a silly example, but there are lots of small items you will not choose to include in your PM program for this same reason.

If you can repair or replace a piece of equipment as easily and cheaply as you can PM it, then a "run-to-failure" strategy will save time and money in the long run.

Emergency Maintenance

Emergency maintenance is the more urgent sibling of corrective maintenance. While corrective maintenance needs to be done timely, emergency maintenance needs to be done immediately. Emergency maintenance can include gas leaks, broken water pipes, roof leaks, broken windows, backed up sewer lines, no heat calls in winter, no cooling calls in summer, snow and ice removal, or any maintenance issue that puts your facility or people at risk of further harm.

Emergency maintenance is the most disruptive type of maintenance to a well planned PM schedule. Emergency maintenance is the reason I recommend scheduling PM tasks for particular weeks and not for particular days. We have all had days when "all hell breaks loose" at work. If we had PM tasks assigned for that particular day, those tasks may go undone. If we have PM tasks scheduled for any time during that week, there is enough flexibility for your PM program to recover.

Safety, health, and environmental compliance items usually make up a large percentage of emergency maintenance. These items, sometimes referred to as the acronym SHE, can often be prevented through proper PM but need to be attended to immediately when there is a problem that could potentially harm building occupants, the public, or the environment.

Emergency maintenance can have the largest cost per job of any work you do. These maintenance projects often require keeping maintenance staff on-site after hours or calling staff back in the middle of the night. It may also require the services of outside contractors who can charge staggeringly high fees for emergency response. In addition, there's usually not time to get competing price quotes for emergency work which can result in higher costs.

Call-back Maintenance

The category of maintenance that I dislike most is when I have to send one of my maintenance people back to do a repair again. Fortunately this does not happen often.

After making repairs, front-line maintenance personnel should be expected to do two things. First, verify the original problem is gone. This means to not only replace the obviously defective part but to also operate the equipment after the repair is made to verify that it is operating properly. The second expectation is that they spend a few minutes inspecting the rest of the equipment and looking for other problems that might arise. Maintenance personnel need to always be on the lookout for anything that might become a problem later. An attitude of continual improvement is invaluable and is what PM is all about.

PREVENTIVE MAINTENANCE TRUISM #1
You must have an attitude of continual improvement for PM to be successful.

PM should not only happen at the scheduled time. Every time maintenance personnel remove the cover on a piece of machinery to make a repair, they should consider themselves, not only repair technicians, but also PM mechanics looking for and correcting other problems.

Call backs make a maintenance department look incompetent. A mechanic with a large number of call backs may be incompetent.

The flowchart in Figure 1-3 shows the relationship between the different types of maintenance.

PM PLANNING BEGINS WITH DESIGN

Most of us did not have the opportunity to help with planning our buildings. Most of us maintain buildings that were built long before our tenure. The best maintenance planning begins with a building's design.

Many times, building designers neglect considering future maintenance needs in favor of creating aesthetically pleasing spaces or keeping construction costs down. We've all worked on air handlers above drop ceilings, been in plumbing chases that would make rats claustrophobic or had to use a hand mirror to see inside some piece of equipment that needed repairing. If you are fortunate enough to ever have the opportunity to help in the design process of a new building, emphasize the importance of considering future maintenance needs.

Like most of us, you are probably working in buildings that were built without a lot of consideration for those of us who would someday have to take care of the facility. Even in this situation, there are things you can do to make your building easy to care for as possible.

By standardizing items such as plumbing fixtures, emergency lights, and door hardware you can reduce your repair parts inventory and reduce the learning curve on new equipment. When ordering parts or replacement equipment, try to order exactly what you already have in place and try to keep everything the same. Having to stock parts for 30 different brands and models of faucets will take up a storage cabinet. If all the faucets in your facility are the same, a small drawer of parts is all you need.

Taking the time to create easier access to equipment that requires frequent service will save you time in the long run. This may mean installing access doors in walls or ceilings, installing ladders or roof hatches, or even relocating equipment to a more suitable location. The hours spent making these modifications can save many more hours in the accumulated minutes required to get to things that are hard to reach.

Sometimes we can find ways to make equipment easier to service. Difficult to reach grease fittings can be relocated using grease port extension kits. Access doors can be cut into equipment to allow access to areas that need frequent service. Sight glasses or Plexiglas or Lexan windows can be installed to allow inspection of the inside of all sorts of equipment without requiring the removal of any machinery covers. Remember that the easier it is to perform PM, the more likely it is that PM will be done.

Whenever building alterations are being planned or major equipment is being purchased, the maintenance department should be included in the process to ensure that future maintenance needs are being properly considered.

CHAPTER 1 SUMMARY

- Preventive maintenance includes all of the work we do to keep our building components and equipment operating in their original condition.
- **Preventive maintenance** is a scheduled program of regular inspections, adjustments, lubrication, or replacement of worn or failing parts in order to maintain an asset's function, and efficiency.
- PM tasks include greasing and oiling bearings, changing filters, changing oil, group re-lamping light fixtures, inspecting drive belts, and many other simple tasks.
- Predictive maintenance (PdM) depends on the condition of an asset to decide if it is the right time to perform maintenance. While preventive maintenance (PM) schedules are time based.
- PM does not require condition monitoring which can be costly.
- PdM does not schedule unnecessary maintenance which can also be costly.
- Preventive maintenance is an important part of facilities maintenance because:

- It extends equipment service life
- It reduces equipment break downs and emergencies
- It saves money by extending service life and maintaining equipment efficiency
- It improves the experience of building occupants
- It makes the work of the maintenance department more manageable

- All maintenance activities that are not PM will fall into the categories of
  - Predictive Maintenance
  - Corrective Maintenance
  - Deferred Maintenance
  - Emergency Maintenance
  - Call Backs

- When an effective PM program is implemented, the focus of the maintenance department will shift from dealing with breakdowns and emergencies to planning and following a schedule of preventive maintenance tasks.

TRUISM #1
You must have an attitude of continual improvement for PM to be successful.

- Building design and prior planning can have a large impact on the maintainability of a building. Considerations such as equipment access and standardizing of repair parts can make maintaining a building much easier. The easier it is to maintain a building, the better the chances are it will be done properly.

Chapter 2

The Economics of Preventive Maintenance

If you have decided that your facility needs a preventative maintenance program, you will probably need to convince someone in your organization your cause is worthy of funding. The savings a PM program will provide will always be bigger than the costs of implementing the program but there will still be costs. There will be a cost for labor to do the PM work, which may include overtime until the program has an chance to start working. You may decide to invest in some CMMS computer software to manage your new PM program. Some new tools and equipment may be needed to take on some new types of work. Turning your existing maintenance staff into successful preventive maintenance mechanics will probably require some additional training. All of these things cost money and need funding.

It is important for the maintenance manager to remember no matter what business your company is in, its primary reason to be in business is to make money. Even if you work for a government organization or non-profit, money is still going to be the primary driving factor in how your organization operates.

Businesses make capital investment decisions based on the potential *return on investment* (ROI) of an opportunity. ROI is a comparison of the money that can be made (or lost) from an investment to the cost of the investment. An investment with high initial costs and low expected return would be a poor investment with a low ROI. While an investment with a low initial cost and high expected return would be a wise investment with a high ROI. Most good business investments have an ROI in the range of 5% to 15%. This means that for every $100 invested, the expected gross return would typically be between $105 and $115. Businesses are always looking for the opportunities with the highest ROI.

Businesses like to make decisions based on ROI, but it can be difficult to calculate the ROI for the activities of the facilities maintenance department. It is nearly impossible to predict how much revenue can be made by spending money on a project such as replacing the lobby carpet. It is assumed that a nice lobby will have a positive impact on the people visiting your company and therefore lead to more sales and increased employee productivity, but it is not possible to really know how many more sales or how much more productivity. Therefore, much of facilities budgeting is guesswork and assumptions. Typically, we know what we spent last year so we try to spend about the same this year. No one really considers how an increase or decrease in maintenance funding would affect a businesses profits.

It can be hard to convince senior management to invest in facilities projects, because so much of the work we do contributes to the condition of the workplace but has no measurable return. Therefore, maintenance is usually seen as a cost not an investment. Most senior financial people see maintenance as a necessary evil that does not contribute to the bottom line.

However, preventive maintenance is one maintenance activity of the maintenance department that can be shown to contribute to the bottom line. To know the ROI of a PM activity, we need to know two things: The initial cost or investment and the return, in our case money saved. We know the initial investment because we can predict the man hours and materials needed to perform PM. And, since we can predict how much money we will save by extending equipment life, saving energy, and reducing repairs, we know the return on our investment. Keep in mine that a dollar saved by PM contributes as much to the bottom line as a dollar earned in other business activities. A dollar is a dollar.

PREVENTIVE MAINTENANCE TRUISM #2
A dollar saved through PM is as good as a dollar earned through any other business activity.

THE JONES-LANG-LASALLE REPORT

While PM in various forms has been a part of facilities maintenance for decades, very little research has been done on determining the real-world ROI of PM. It has always been assumed that PM has a positive impact due to increased equipment life, energy savings, improved tenant satisfaction, and reduced breakdowns. Every maintenance manager can cite an example of how PM has impacted their business, but very little in the way of hard numbers has been available.

Many organizations have developed industry benchmarks of how much money "should" be spent on PM. Unfortunately, these are based on surveys of what businesses are spending, not on any measure of what spending level has the greatest return on the investment.

In 2000, Wei Lin Koo and Tracy Van Hoy, PE., both working for the real estate management firm Jones Lang LaSalle published a research paper titled "Determining the Economic Value of Preventive Maintenance." By analyzing maintenance costs across a wide range of facilities, Koo and Van Hoy were able to determine average ROIs for several different types of capital equipment.

Their project looked at 25 years of maintenance and equipment replacement costs for roughly one million square feet of facilities across several industries. They found what many had believed to be true. They discovered that investing in good PM programs had huge financial returns. Figure 2-1 shows the average ROI for PM for several different types of capital equipment.

From this chart, it is clear to see that investing capital in the tools and man power to perform PM can have enormous returns. We said earlier that typical ROIs for successful business investments tend to be in the neighborhood of 5-15%. ROIs of 500%, 1000%, or 2000% are simply unheard of. I do not know of any other investment that can provide as high an ROI as a good PM program. These high ROIs are possible because the cost of replacement of capital equipment is so high and the cost of maintaining them is so low. As an example, if we were to completely neglect a small 1Hp circulating pump, we will probably be replacing the pump at a cost of several hundred dollars in three to five years. With less than ten minutes labor and a few pennies worth of grease every six months, that same pump can reasonably be expected to last 20 years. It is easy to see how a small investment in labor and materials to grease a pump bearing can prevent the cost of replacing the same pump 3 or 4 times over 20 years.

In future chapters of this book we will be discussing what equipment we will include in our preventive maintenance program and what equipment not to include. The possible ROIs of the different types of equipment will be a large factor in making that decision.

SELLING PM TO MANAGEMENT

If PM is such a sound investment, why do so few facilities have comprehensive PM programs? There are probably three reasons for this.

Capital Equipment Type	ROI
Air Compressor	230%
Air Handler	100%
Boilers	850%
Centrifugal Chillers	1,100%
Reciprocating Chillers	400%
Cooling Towers	550%
Condensers (air cooled)	1,050%
DX Units	1,800%
Fire Detection Systems	10%
Centrifugal Pumps	2,300%
Fire Pumps	50%
Switch Gear	700%
Parking Lots	900%
Roofs	350%

PM's Return on Investment as determined by Koo and Van Hoy, 2000.

Figure 2-1

First, preventive maintenance has not fully caught on in the facilities maintenance field. Industry and manufacturing depend entirely on their machinery and equipment to operate. Industry and manufacturing businesses are crippled every time a machine breaks down. Failing equipment is a daily reminder to plant management that PM is absolutely necessary.

In the field of facilities maintenance, a single broken piece of equipment is not usually crippling to the operation of the business. If one air compressor in a building is off line for a few hours or days, business can

Figure 2-2. Money spent on PM for centrifugal chillers has a return on investment of over 1,000%.

usually continue without too much inconvenience. Although the cost of a breakdown is more expensive than the cost of PM, the urgency to invest in PM is just not there.

PM is often low on the list of management priorities for a second reason. PM addresses problems that might happen, but these problems are not guaranteed to happen. There is a reluctance to spend money on maybes, even if the long-term benefits are obvious. Businesses have limited resources and they already have guaranteed expenses that cannot be ignored. Weekly payroll will happen every week. The electric bill will need to be paid every month. Taxes will be due each quarter. But the cost of a major repair is not as certain.

Even when management understands the value of an effective PM program, there can be resistance to making the necessary small investment to prevent problems that may or may not happen. It is human nature to assume that since nothing catastrophic happened last month we can probably get by for another month or two without PM. Unfortunately, it is often the business-crippling equipment failure that pushes PM to the top of management's priority list. When this happens, resist the temptation to say "I told you so."

PREVENTIVE MAINTENANCE TRUISM #3
You cannot save money by skimping on PM

The third reason management is sometimes resistant to invest in PM is that PM does not provide much benefit in the very short term and a lot of the benefits of PM are hard to quantify. When a business invests in opening a new branch, management can see, touch, and feel the new building. But when they invest in PM, they cannot see what they spent their money on. They see nothing. Of course, in this instance, seeing nothing is a good thing. Nothing is the best thing a maintenance department can have. In maintenance, the alternative to nothing is problems. However, it is human nature to want to see what we buy. PM does not have anything to show, at least in the short term.

In the long term, the benefits will be seen in reduced maintenance costs and increased maintenance efficiency. It can be hard to convince a lot of people to think in terms of long-term benefits.

Most companies also already have a backlog of things to invest money in that could save them money. Purchasing more efficient heating and cooling equipment to reduce electrical consumption, converting to more efficient lighting, adding insulation to the building's envelope to save heating costs, installing window tinting to reduce thermal heat gain, purchasing new photocopiers that will require fewer service calls, and installing new computer or phone systems that will increase productivity may already be on a companies to do list. Preventive maintenance might be just another money-saving project awaiting funding.

If the money to start your PM program is not available right now, you may be able to start small by excluding equipment from your PM program expecting to add equipment later. Knowing the expected ROI of various types of equipment will help you to make the decision on which pieces of equipment to add. ROI should not be the only consideration if you are forced to limit the equipment to PM. You will also need to consider what equipment is causing the most disruptions to your department and which equipment is having the biggest impact on building occupants. By choosing to PM the equipment that is taking up the biggest part of the maintenance department's time with repairs, it may be possible to reduce those repairs and find more time to PM more equipment.

PM AS A BUDGETING TOOL

In most organizations, near the year's end, the facilities department is asked to submit a budget for the following year so corporate funds can be allocated as needed. Many facilities departments just add a few percentage points to each line item of this year's budget and assume that costs will remain the same with this small adjustment for inflation.

If we have been performing preventive maintenance to all of our capital equipment and keeping accurate records of maintenance and repairs, we will be much more prepared to develop an accurate budget of expected expenditures. If we have been inspecting our capital equipment according to our PM program, we will be aware of what equipment needs replacing, what repairs are likely, and can budget more effectively.

In an effort to control costs, facilities departments often attempt to "limp along" old equipment until it reaches the point that no more repairs can be done. With good repair and inspection records, we will know when the cost of repairs was exceeding the cost of replacement and would be less likely to continue to waste money when replacement would be more cost effective. We can plan to replace equipment that is nearing its end of use and is starting to cost excessive money in repairs and maintenance.

If we are not inspecting our equipment regularly as a part of a PM program and are not keeping accurate records, budgeting becomes a guessing game with lots of unexpected expenses that were not accounted for. We will discuss the details of record keeping later in Chapter 3.

THE LONG-RANGE FACILITIES PLAN

It can have lots of different names such as "long-range planning," "five-year asset improvement plan," "three-year plan" or the "comprehensive maintenance plan" but most organizations try to do some broad planning at least three to five years into the future. The purpose of such planning is two-fold. First, to make sure that there is some continuity to the goals and direction a company is taking rather than changing direction every year. And second, to anticipate future costs so there are no budget surprises.

Most medium and large companies see the building and property as an important part of its long-term planning and include facilities needs in this planning process. If it is not already part of your corporate culture, you should try to get your department's long range planning included in the process.

If you have been keeping records of repairs, know your maintenance costs, and have been doing regular PM inspections; you should have an idea of when a particular piece of equipment is nearing the end of its useful life. Having future replacement costs of major equipment as part of the long range plan will help to soften the blow when you need money to make the replacement. The costs of any deferred maintenance projects should also be considered for inclusion in the long range financial plan.

Very large and expensive capital projects such as new roofs, repaving parking lots, and building wide upgrades of electrical or plumbing systems are sometimes only possible if financial plans were made years in advance. For particularly costly improvements, budgeting only one year in advance is often not enough time for the necessary funds to be found.

An extended outlook facilities plan should include all of the corrective maintenance (equipment repairs or replacement) and capital improvement anticipated for the next several years. In Chapter 3, we will discuss developing an inventory of all of a buildings capital assets or equipment. This list of capital assets should be reviewed and each item's repair and inspection history reviewed to determine what repairs or replacements will likely be needed in the future.

A backlog of maintenance work orders (maintenance requests) and any maintenance tasks that were deferred should also be reviewed for inclusion in the long range plan. Minor maintenance work can be considered the cost of catching up and there may be maintenance requests that are large enough in scope and cost that they warrant inclusion as a capital project. A maintenance request to construct walls to divide a large meeting room into several offices could be one example of work that should be included in long range planning unless budget money and labor is available to do the work now.

THE REPAIR OR REPLACEMENT DECISION

The decision to repair or replace a piece of equipment is one that is made every day by most maintenance managers. When faced with this decision, experienced facilities people usually rely on our experience working with similar types of equipment. We will look over a broken air handler, irrigation pump, or compressor and see there may be other failing components and that the machine is either in relatively new condition or nearly worn out and make a decision based on our gut feeling. It is not very scientific but it is the method most of us use. When particularly large and expensive equipment is involved or when expensive repairs are needed, it would be nice to have a way to make the repair vs. replace decision that relies on something more accurate than old fashioned intuition.

We have already discussed several reasons for maintaining good equipment history records. When determining if it will be more cost effective to repair or replace a particular piece of equipment, these records will provide the information to make the right decision and the evidence to convince others that the expenditure is justified.

There are two economic reasons to decide to replace a piece of equipment that has stopped working. The first is the cost of maintaining and operating the equipment will exceed the cost of replacement. The second is other similar pieces of equipment reached the end of their useful life at approximately the age of this one.

The first scenario, repair costs will exceed replacement costs. This is easy to understand. If a RTU (rooftop air-conditioning unit) will cost $9,000 to replace and the expected cost of repairs is $10,000, it makes sense to replace rather than repair. But we have to know the time frame for repairs before we can make the right decision. If the cost of repairs exceeds the cost of replacement this year, then the decision is easy. Both costs would

Figure 2-3. Making a repair-vs.-replace decision on large equipment, based on intuition, is a sure way to waste money.

come out of this year's budget and replacement will have less impact on this year's budget than repair so we replace.

However, if we know that replacement costs today would be $9,000 and repair costs will be $2,000 each year for a total cost of $10,000 over five years, it becomes less obvious whether repair or replacement today is the right choice. Repair versus replacement decisions are often subjective. Factors such as expected useful life, available budget, and others concerns can make the repair versus replace waters murky.

The second economic reason to replace a piece of equipment is other similar pieces of equipment have reached the end of their useful life at the age of the piece in question. For example, assume you work in a hotel that has individual package terminal air conditioners (PTACs) in each hotel room. PTACs are those AC and heat units that fit in a sleeve through an exterior wall and allow each guest to adjust the temperature in their own room. You have a hundred or more of these PTACs in your hotel, all of the same model and brand. You know from your repair records that 70% of these units develop leaks in the evaporator and condenser coils by their eleventh year of service and the cost coil replacement exceeds the cost of a new unit.

If one of these PTAC units was sent down to your shop with a bad compressor, you will need to decide either to repair the unit by installing a new compressor or if it would be more responsible to replace the entire PTAC with a new one. The replacement compressor costs roughly half the cost of a new unit and this particular PTAC is 12 years old. Since we know that only a few units survive past 11 years, this unit does not have much service life left. If we install a new compressor on this unit, we are essentially wasting a compressor since we will likely end up disposing of the old unit with newly installed compressor in the very near future.

What if the PTAC needing repair was only eight years old, or only six years old? Would replacement or repair be the proper option? Fortunately, all of these questions can be answered with a simple math trick.

The trick is to figure out what the average cost of each option will be over the life of the equipment. Trying to compare $10,000 in repairs over five years with a one-time replacement cost of $9,000 is impossible. One time costs and costs over several years are not the same thing. We need to find a common unit of time that we can use for both options. As the old saying goes, "compare apples to apples, not apples to oranges." The way we do this is to figure out the equivalent annual cost or EAC of each option. After we have both options converted to the same unit of time, the comparison is easy. EAC is the total cost of an item divided by the number of years an item will be used.

As a simple example, if we spend $1 Million on a new roof and expect that roof to last 20 years, the equivalent annual cost (EAC) for the roof would be

$1M/20 years or $50,000 per year.

In another example, if we believe that replacing a $8,000 oil burner will give us another 5 years of service life from a boiler, the EAC of this repair would be

$8,000/5 years, giving an EAC of $1,600 per year.

Let's look at how we can use the concept of EAC to decide whether to repair or replace the RTU and PTAC in the examples just given.

EXAMPLE 1—REPLACE OR REPAIR RTU (rooftop unit)
Estimated cost of repair: $10,000 over the next 5 years.
Estimated cost of replacement: $9,000 one time
Estimated service life (how long a new unit will last): 12 years.

To compare the cost of replacement with the cost of repair for a piece of equipment, we will need to add up all of our expenses that we can expect to occur over a specified period of time. We will also need to decide what that period of time will be. We need to decide on a period of time first.

If we replace the unit with a new one, we can expect the new RTU to have a service life of 12 years. That is a good average rule of thumb for this type of equipment. If we continue to repair the old RTU, we believe that we can get 5 more years out of the unit.

Option 1 repair RTU unit
Total cost of repairs: $10,000
Total expected years of service: 5
$10,000/5 years = $2,000 per year

Option 2 replace RTU unit
Total cost of replacement: $9,000
Total years of service: 12
$9,000/12 years = $750, per year

It should be easy to see, from this simplified problem, that option 2 (replacement) is less expensive in the long run then option 1 (repair). Replacing the unit is therefore the "right" answer.

It should be noted that although option 2 above is the "right" answer, if there is no money in the budget this year, an organization may opt to "waste" the repair costs for a couple of years until the replacement can budgeted. Sometimes even if we know what the right option is, the real world forces us to make another choice.

Let's look at the how we would make the repair vs. replace decision on the PTAC mentioned earlier.

EXAMPLE 2—REPLACE OR REPAIR PTAC (package terminal air conditioner)
Estimated cost of repair: $450 for new compressor
Estimated cost of new PTAC: $900 one time
Estimated service life if repaired: 1 year (our best guess based on other PTACs)
Estimated service life is replaced: 11 years

Option 1 repair PTAC unit
Total cost of repairs: $450
Total years of service: 1
$450/1 year = $450 per year

Option 2 replace PTAC unit
Total cost of replacement: $900
Total years of service: 11
$900/11 years = $81.82 per year

We can see that replacing the unit is much less expensive over the life cycle of the unit. We would therefore choose to replace the PTAC instead of making the repair.

Many books have been written about engineering economics. The examples chosen above are intentionally simplified because this is not a book about engineering economics. This is a book about preventive maintenance and making the right repair or replace decision is important to those of us that do PM. As part of our PM inspections, we are going to find problems with older pieces of equipment and are going to have to make these types of decisions.

These two examples only consider the costs of repairing the existing equipment or installing replacement equipment. In the real world, ALL of the costs associated with each option would need to be considered. Some of these costs would be:

- The cost of maintaining (PM and repairs) a new piece of equipment for each year of service
- Down time costs while repairs are made each time they are made
- Down time costs while a replacement is orders
- Equipment disposal costs
- Depreciation
- Salvage value
- Available utility or government rebate programs for equipment upgrades
- The costs of differences in energy efficiency over time
- The cost of inflation

All of these costs can be considered in the same way as the costs in the above example. Knowing how to intelligently make the decision to repair or replace can have a large impact on the bottom line and therefore on your department budget. Making the wrong decision on a $9,000 RTU or on a $900 PTAC most likely will not break the bank. Many facilities projects reach into the millions of dollars. In these instances, a wrong decision based only on intuition or a gut feeling can be very expensive.

THE USEFUL LIFE OF BUILDING SYSTEMS

Budgeting, long-range planning, and making repair vs. replacement decisions all depend on knowing the useful service life of different types of equipment. There is no way to know how many years a particular piece of equipment will last, anymore than you can predict exactly when a light bulb will burn out. However, there are rules of thumb about expected equipment service life based on years of observation. These rules of thumb will provide some direction in predicting the useful life of different types of equipment.

Environmental factors, hours of operation, and maintenance history will have an impact on the service life of machinery. Salty sea air, exposure to the sun and rain, and use under heavy loads will reduce the service life of most equipment. Conversely, like the little old lady's car that only gets driven to church on Sundays, light use and protection from the elements will usually result in an increase in service life. These numbers are only a guideline. Many pieces of equipment, if maintained well, can last years beyond their expected service life while other pieces may fail earlier.

ESTIMATING PM COSTS

As mentioned earlier, a preventive maintenance program should cost nothing in the long run. In fact, a PM program should save money for an organization.

Although PM will save money in the long run, there are immediate costs associated with performing the work of PM. While these costs are more than offset by all of the associated savings discussed in Chapter 1, they are still real costs and need to be accounted for.

If PM is a new concept to your organization, it will be met with some caution on the part of those who hold the purse strings. The first question asked will most likely be "How much will this cost?" not "How much will this save?"

Expected Service Life of Selected Equipment

Item	Years	Item	Years
Air Conditioning		**Plumbing**	
Central Air	10-15	Black Iron Pipe	20-25
Central Chiller	20-25	Cast Iron Pipe	30-40
Cooling Towers	15-25	Circulator Pump	20-25
Window units	10	Copper Pipe	25-30
Electrical		Drinking Fountains	10-15
Building Wiring	20-25	Fire Sprinkler	30
Exterior Wiring	15-25	Galvanized Pipe	25-30
Lighting	20	Plastic Pipe	25
Service Equipment	20-30	Water Heaters	10
Exterior		**Roofing**	
Asphalt Paving	15	Asphalt Shingle	15-20
Chain Link Fence	20-30	Built Up Roof (BUR)	15-20
Lawn Irrigation	10-15	Modified Bitumen (MBR)	15-20
Sidewalks	20-25	Rubber (EPDM)	20-30
Heating		Standing Seam	15-25
Boiler or Furnace	20-30	**Structural**	
Electric Unit Heater	15-20	Concrete/Masonry	45-75
Gas Unit Heater	20-25	Steel Frame	35-50
Hot Water Heater	20-25	Wood Frame	25-40

Figure 2-4. Expected service life of equipment

A preventive maintenance program will eventually reduce the emergencies that your maintenance department is dealing with. In the long run, the extra work PM requires will not increase your workload since unplanned maintenance work will be reduced. Your work will simply shift to more planned activities and less emergencies. In fact, there will be time to tackle the work you have probably put on hold due to a lack of time. However, in the very short term performing PM will require extra man hours, either in overtime or an increase in staff.

Many organizations are convinced that there is not enough money, time, or staff available to take on a PM program. Of course, these organizations will somehow find the money, the time, and hire the staff to repair what was not maintained. The familiar adage "Pay me now or pay me later" should become "Pay me a little now or pay me a lot more later."

Starting PM will require additional work until your PM program starts to show some benefits. Ideally, you will be able to either pay overtime or hire temporary help. If this is not possible, do not give up on PM. Motivate your staff; explain how much easier work will be if they can get some extra work done during the day for only 3 months. Start slowly, focusing on the PM tasks that will have the highest impact on your workload and then use the extra time saved by those tasks to start more tasks. In the perfect world, we would all have the money and labor we need to start our PM program all at once. In the real world of budget and time restraints, sometimes it is necessary to take baby steps.

The second area of PM costs, after labor, will be material costs. Material costs will usually be much less than the labor costs since the material involved in PM are usually very inexpensive and used in small quantities. These include oil, grease, and other lubricants, replacement filters, parking lot sealer, paint, caulk, and the occasional replacement part if a defective part is found during an inspection.

The third area of PM costs will be the one time cost of modifying equipment or the building to make PM easier. Equipment that is difficult to PM will end up being skipped if it is not altered in some way to make PM easier. Adding access doors to air handlers, installing ladders to reach inaccessible roofs, and installing grease-port extension-lines to move grease fittings within reach will all increase the likelihood that someone will take the time to do PM instead of skipping a difficult piece of equipment.

The only way to answer the question, "How much will this cost?," is to add up the costs of labor, materials, and equipment modifications. In Chapter 3 we will create a PM schedule and calculate all of the labor hours required for each task. The cost of materials can be estimated fairly quickly after you have done an equipment inventory and know what types of service each piece of equipment will need. The last area of costs, equipment modifications, will be harder to determine until after your PM program has started and your staff tells you the difficulty they have in performing each task. However, these modifications can be done slowly, stretched out over enough time, as your budget allows. Do not neglect these modifications; my experience is that equipment that is difficult to PM does not get maintained properly. A facilities department mindset of continual improvement will help to make your PM program successful.

CHAPTER 2 SUMMARY

- Business decisions are made according to the anticipated return on investment (ROI). Preventive maintenance (PM) offers some of the highest ROIs available to business in the form of cost savings.
- The cost savings created by an effective PM program are primarily due to extended equipment service life and maintained efficiency.

PREVENTIVE MAINTENANCE TRUISM #2
A dollar saved through PM is as good as a dollar earned through any other business activity

- The repair and inspection data collected during PM activities will be valuable in helping to predict future expenses for developing budgets and long-term planning for your organization.
- When deciding if it will be more economical to repair or replace a particular building component, the Equivalent Annual Costs (EAC) of each option needs to be calculated. Once the EAC of each option is known it is easy to compare the cost of replacement with the cost of repair.
- Another important piece of information needed for accurate budgeting, long range planning, and making repair vs. replacement decisions is the expected service life of different types of building components.
- Although PM will offer cost savings in the long term, there will be short term start up costs when beginning a PM program. These costs will come from labor, materials, and necessary modifications which will make equipment and building components accessible for PM.

Chapter 3

Getting to Work–Setting up a PM program

There are many reasons that so many facilities don't have preventive maintenance programs in place. Most reasons for not having a PM program are about limited time and money. You may have heard some of these reasons in your organization.

TOO BUSY FOR PREVENTIVE MAINTENANCE

If your department is so swamped with emergency repairs that you can't find time for PM, you desperately need to find time for PM. A good preventive maintenance program will significantly reduce the breakdowns and other emergencies that are keeping you so busy. A good way to start a PM program when you are over extended is to identify the ten pieces of equipment that require the most man hours in repairs and start with these ten things. Once properly maintained, the emergency repairs on these top ten should slow down and they will no longer be your top ten time wasters. Use the tiny bit of extra time you've saved in repairs to these items to add the next top ten pieces of equipment to your program. In sort order, you will have a comprehensive program and will no longer be chasing your tail.

NOT ENOUGH MAINTENANCE STAFF

This may be true. You may legitimately understaffed for the size and age of your facility. However, failing to perform PM is not a good strategy for making your small staff as effective as possible. Dealing with emergencies from neglected equipment requires many more man hours than is required to perform proper maintenance. These are hours you can't afford to waste.

CAN'T AFFORD TO HIRE SKILLED STAFF FOR SPECIALIZED TASKS

This is an idea that is common in many facilities. A familiar example of this is when a roof has a leak. In-house maintenance staff are used to make roof repairs even if they don't have the necessary knowledge and skills to do it right. Slopping roof patch on the hole or separated seam stops the leak for a few weeks. The result is that instead of spending the money for a proper repair once, you end up spending money over and over again to repair the roof. Include the cost of replacing ceiling tiles, carpets, and other things that get wet inside the building and it's easy to see that saving money on skills rarely saves anything.

DON'T NEED TO PM SOMETHING THAT'S ALREADY PAST ITS USEFUL SERVICE LIFE

Whatever the equipment is, if you haven't already replaced it, you probably are going to need to continue using it. If replacement equipment is already ordered and scheduled for delivery in the next few days, then skip the PM. Otherwise, continue to PM the equipment. PM on a brand new piece of equipment doesn't make a lot of difference in the equipment's performance. PM on older equipment makes a big difference. Failure is more likely as equipment ages. With proper PM, its useful life might be extended beyond your expectations.

STARTING TO DO PREVENTIVE MAINTENANCE

Creating an effective preventive maintenance program for your facility will require time and dedication on the part of you and your maintenance staff. Starting a PM program from scratch is a big project that will require time spent in addition to your regular maintenance work. Doing this work can be discouraging since the benefits of a good PM program take months to materialize. If you can stay the course and successfully follow your PM program for the first 3 months, the benefits will be dramatic, both in the long-term costs of operating your property and in the way the maintenance department manages its workload.

If you are bombarded with a constant barrage of emergencies preventing you from finding the time to set up a PM program, you desperately need to find the time to get your building under control. The work required to create a new PM program will most likely fall on the maintenance manager's shoulders.

After the program is created and implemented, the extra work begins for the maintenance mechanics or technicians. Maintenance department personnel will essentially be doing two jobs during the first 3 months that a PM program is in place. They will need to continue to do their usual maintenance repairs and still find time to perform several preventive maintenance tasks each week.

There are as many variations on PM programs as there are facilities. Some PM programs are good and some are not. The following 4 step process will help you to develop a PM program that is effective and manageable. If you follow these 4 steps, you will have a PM program that will improve the condition of your facility, improve the experience of your building users, and improve the way the maintenance department works every day.

The 4 steps are (1) determine your organization's goals with respect to PM, (2) inventory your building's equipment and use your organization's goals to decide which equipment to include in the program. (3) figure out what type of preventive maintenance each piece of equipment needs and how often. (4) put all of this information into a calendar that will be your preventive maintenance schedule.

Step 1—Determining Your Organization's Goals

All organizations are different. Different organizations have different standards and expectations from the facilities and maintenance department. Therefore, not all PM programs will be the same.

The PM tasks that care for mechanical equipment will be fairly universal from building to building. If the same model of boiler is installed in a hospital, school, or hotel, the manufacturer's maintenance recommendations will be the same. The same is true for nearly all mechanical equipment. The PM procedures and schedules for mechanical equipment will be identical regardless of where they are installed.

By contrast, let's consider those PM tasks which affect only the aesthetics of a building, such as painting, wallpapering, buffing floors, and changing stained or dirty ceiling tiles. The PM schedule for these tasks will vary widely from facility to facility. This is because different facilities have different standards concerning appearance. A warehousing and shipping facility will have less emphasis on aesthetics than you would expect at a hotel. The PM schedule for a warehouse might include painting the lobby every five years, while paint in hotel guest rooms is touched up every three months. Each industry has its own priorities that must be taken into consideration when setting up PM schedules. A successful PM program for one industry will not work well in another.

Not only do PM schedules vary across industries but the expected service life of carpets, wallpaper, and paint will also vary from industry to industry. A slightly worn carpet may be perfectly acceptable for use in a factory break, room but would have been thrown out years earlier if it were in the main lobby of a financial investment firm. The rules of what is aesthetically acceptable are hard to define.

Most organizations place higher priority on the appearance of public places such as lobbies or building facades and place lower priorities on places only used or seen by employees. Stock rooms are rarely as attractive as the adjoining retail space. When setting up your PM program, you will need to consider the usage of each space. All spaces do not require the same PM frequency since all areas do not have the same standards for appearance.

In order to create an effective PM program, you will need to understand your organization's goals with respect to its appearance. These goals are often hard to define, rarely written down, and different managers will likely have different opinions on the matter. If you do not feel confident that you have clear direction in this area, ask. Ask those who have the authority to spend the necessary money, to look at your building to determine if the present condition is acceptable. If not, ask if they are willing to invest the resources to bring the building up to standards.

Step 2—Inventory Capital Equipment and Decide What to PM

Before you can PM any piece of equipment, you will need to decide what equipment to include in your preventive maintenance program. Each building component and each piece of capital equipment in your building will need to be considered to determine if it should be included.

Blueprints or construction documents for your facility will be a good resource when developing your equipment list. The floor plans and mechanical drawing pages will help you to locate equipment which may be hidden out of view during your walk through. Air handlers, shut off valves, and lighting timers are often hidden above suspended ceilings and other concealed spaces. A set of prints will show you where to look for these items.

The blueprint's equipment schedules of HVAC equipment, electrical panels, and plumbing fixtures will provide manufacturers, model numbers, voltages, and lots of other important information about each piece of equipment. Since equipment may have been added, removed, or replaced since initial construction, a thorough walk-through of the facility will also be required to verify the information found on blueprints.

Both mechanical equipment and building areas will need to be considered for inclusion in your PM program. Chapter 11 discusses the actual PM procedures and PM frequency for different types of mechanical equipment and will be a useful resource to make sure no piece of equipment is overlooked. A walkthrough and review of your buildings floor plans will ensure that all necessary building equipment and areas are included.

There are several reasons to include equipment in a PM program. As we've already discussed, the most common reason to include equipment in the PM program is to extend its service life, reduce breakdowns, and ultimately save money. There are also other reasons to PM equipment. Some equipment must have regular service or inspections performed to comply

Figure 3-1. Some equipment is PMd for safety and not necessarily to extend equipment life.

with government regulations. Examples of this would be annual service to fire extinguishers, annual boiler inspections, and monthly testing of emergency lighting.

Some PM tasks may be included because they affect public safety. Safety inspections of playgrounds and inspections of asbestos containing building materials do not extend equipment life and may not be required by regulations. However, they should be included for safety and liability's sake. Chapter 11 has a section dedicated to items needing scheduled PM according to codes and regulations.

The 80-20 rule says that 80% of the results come from 20% of the causes. The 80-20 rule seems to apply to a surprisingly large number of situations. For most businesses, the top 20% of customers account for approximately 80% of sales. Roughly 80% of income goes to 20% of the population. We tend to drive on the same 20% of local roads, 80% of the time we spend in the car. Most of us spend our free time with the closest 20% of our friends during 80% of our social life.

The numbers may vary a little but the principal that 80% of results come from only 20% of what we do also applies to PM. When selecting equipment to include in your PM program, Roofing and HVAC equipment are the 20% of your building's equipment that will result in 80% of your PM program's success. It is interesting to note that roofing and HVAC systems account for about 17% (very close to 20%) of commercial building construction costs. Don't be surprised if you find that 80% of your PM work is concentrated on this 20% of your building.

The equipment lists in Chapter 11 can be used as guides to help you decide what to include in your PM program during your walkthrough and equipment inventory.

Equipment Data Sheets

During your walkthrough to identify the equipment in your building, you will obviously need to write down what you find. It will be helpful to know not only what type of equipment you have but some other information about the equipment. This information will be used later to determine the frequency and actual PM procedures for each piece of equipment. I prefer to write this information on an equipment data sheet like the one in Figure 3-2. The data collected should include everything needed to research the manufacturer's maintenance instructions including model numbers, serial numbers, dates of manufacture and the manufacturers' addresses and phone numbers.

EQUIPMENT DATA

Property ______________________ Equip. No. ______________________
(type) (number)

Type ______________ Location ______________
Make ______________ Mod. __________ Ser No. __________
Volts ______ Amps ______ Date __________ Price __________

TYPE KEY
A - Air cond.
B - Boilers
C - Compressors
E - Elevators
F - Fans
G - Grounds
K - Kitchen
L - Laundry
M - Motors
O - Other
P - Pumps
R - Refrig.
V - Valves
W - Electronic
X - Emergency

MANUFACTURER - ADDRESS - PHONE	LOCAL - VENDOR - ADDRESS - PHONE

MOTOR DATA

Mfg	Hp	Volts	Amps	R.P.M.	P.H.	Ins Cl	Duty	Frame	Model	Cat. No	Ser No

SPARE PARTS				OTHER DATA
Mfg	Part No.	Model	Cat No.	Fluids - Belts - Fuses - Filters - Etc.

MAINTENANCE REQUIREMENTS

Figure 3-2

If you choose to use an equipment data sheet, I would recommend collecting some additional information. Write down data such as equipment operating voltage and part numbers of those frequently needed items such as filters, drive belts, or fuses. These data sheets then become a valuable resource in the future. Keeping these sheets filed and accessible will make ordering replacement parts easier. You will always have model and serial numbers at your fingertips and will rarely need to make a trip to the roof to get a part number for a broken drive belt.

By using equipment data sheets to record equipment information during your initial inventory, you can be sure all necessary information is collected and you have the information available in a standardized format. This will reduce the need to re-walk the building for missed information and will help when determining what PM tasks need to be done and in setting up your PM calendar later in this chapter. Even if you will be using a computer maintenance management system (CMMS) to manage your preventive maintenance program, equipment data sheets provide an easy to use format to transfer the information into the computer later.

One more note about equipment data sheets: Many maintenance departments use the back of these sheets as a permanent record of equipment repairs. Each time a technician completes a repair, he should fill out the back of the equipment data sheet for that piece of equipment. Having a complete repair history of each and every piece of equipment in your building will be useful in future troubleshooting, making repair vs. replace decisions, and in developing capital budgets.

The equipment data sheet with the repair history should remain in an active file as long as the equipment is in service. When a new piece of equipment is installed, a new equipment data sheet is filled out and added to the file. When an old piece of equipment is retired, its equipment data sheet is archived.

In many organizations, equipment data sheets have given way to computerized equipment inventories. Many CMMS (computerized maintenance management system or software) are capable of developing PM schedules for most types of equipment and can also maintain a repair history on every piece of equipment in the inventory. Whether you choose to set up a computerized PM system or a paper system, the initial equipment inventories are done the same way.

Unique Identifier

Every piece of equipment in your PM program should be given a unique identification number. It may seem logical to identify equipment by its location instead of creating an equipment number but this approach has flaws. Primarily, some pieces of equipment may be moved to other locations. PTACs, those through-the-wall air-conditioning units often used in hotels, are often swapped from room to room when a unit needs repairs. This means the PTAC in room 111 might not be the same PTAC that was in room 111 the last time a repair was made. And when equipment is retired and replaced, we need a way to distinguish the old equipment from the new and start a new repair history.

The most common way facilities department identify equipment is with a two part ID number. The first part of the number identifies the type of equipment, such as A for air conditioner or F for fan. The second part is simply a unique number that lets us identify each air conditioner from the others.

Commonly Used Equipment Identifiers

A—Air conditioner
B—Boilers
C—Compressors
E—Elevators
F—Fans
G—Grounds
K—Kitchen
L—Laundry
M—Motors
O—Other (misc.)
P—Pumps
R—Refrigeration
V—Valves
W—Electronic Equipment
X—Emergency Equipment

By no means are these identifiers set in stone and you will probably have other equipment in your facility that need additional letters to be added. You might even want to include more information in the ID numbering system. In large facilities you might add a third component to identify which area of a building the equipment is located in or a component that identifies specific buildings. In a school district, you might use ID number CP-A04-SH for circulator pump number 04 located in A-hall at the senior high school. With this type of numbering system, everyone in the maintenance department can tell immediately the type of equipment, what building it is in, and where it can be found in the building.

The ID system is completely up to you but plan the system carefully. Your ID system should include all the information you consider important. Your numbering system must allow a unique identifier for each piece of equipment. If you work in multiple buildings, you don't want a pump identified as P-02 in more than one location. Some equipment will already have unique identifiers. For example, electrical panels usually have unique ID numbers attached during construction and boilers usually have unique registration numbers issued by the state.

Instead of doing a walkthrough twice, you can assign a unique equip-

Figure 3-3. Each piece of equipment should be marked with its own unique ID number.

ment identification number to each piece of equipment as you make your equipment inventory. A selection of metal marking paint pens in several bright colors will let you easily write the equipment identification number on each piece of equipment. Make the numbers large, bright, and durable enough to be visible for the equipment's entire service life.

Many organizations maintain asset inventories but only maintain an inventory of equipment which exceeds a preset value, often $1,000 or $2,000. These inventories are maintained for insurance, taxes, and asset control purposes, but not for the purposes of preventive maintenance. While our inventory usually will not include every lock set or window, it should include each piece of equipment that will be part of our PM program. Circulating pumps, ventilation fans, and shut off valves are typically not expensive items but should be inventoried for our purposes since they will be listed on our PM schedule.

Not everything that will be included in your PM program needs its own equipment ID number. For example, public rest rooms should be included in every PM program. Every plumbing fixture and toilet partition in the rest room should be inspected and repairs and adjustments made on a regular basis. Restroom PM will include checking the operation of ventilation fans, lights, faucets, drains, and toilets. You will also inspect wall, ceiling, and floor finishes, checking all fixtures for leaks, and tightening any loose TP rollers, towel bars or robe hooks. Each robe hook or faucet does not need a unique ID number nor do each of these small items need to be included in our equipment inventory. To label and inventory every paper towel dispenser would be too cumbersome and time consuming. Just as a an air handler would be inventoried as one item instead of considering the blower, air coil, filter rack, and cabinet separately; A restroom would also be considered one item with all its components maintained together.

In the rest room example, the rest room does not need a unique ID number since it probably already has one. We probably already call that restroom "the third floor rest room" and can use that title on our PM schedule. Using the room location as the identifier will work fine in this case because, unlike other types of assets, it is unlikely that the third floor rest room will ever be moved anywhere else. Of course this same logic extends to offices, lobbies, guest rooms, classrooms, and lots of other building areas that need to be part of a PM program but don't need to be inventoried.

The rules about what should be included separately and what should be grouped together in a preventive maintenance program aren't always clear. Even if you're considering all the equipment in one hotel room as a single item on your PM schedule, you may still have specific pieces of equipment within the hotel room that you want to inventory separately. For example, you may decide to include hotel room 101 on your PM schedule so that everything in the room is inspected and repairs made once a month. However, you may decide to separately include the PTAC, refrigerator, and television in room 101 in your equipment inventory. The fact that these items can be moved from room to room suggests they should not be considered as part of room 101.

In a commercial kitchens, items such as doors, walls, floors, counters, and lighting can be lumped together as "Kitchen PM," while larger, more complex equipment in the kitchen such as ventilation fans, fire suppression systems, and walk in freezers should be considered separately. In many cases which items to include is a judgment call based on experience and you will likely end up making some adjustments to your system after it is implemented. No two PM systems are the same and that's fine because no two buildings, companies, or maintenance departments are the same.

If you will be using a CMMS (computerized maintenance management system) to manage your PM program, deciding what equipment to lump together and what to consider individually becomes even more important. A typical CMMS will generate paper work orders on the dates particular PM tasks are to be completed. A particular school may have 20 classrooms in one wing with unit ventilators in each classroom. If we are changing unit ventilator filters in that wing on the same evening or even over a one week period, it would make the most sense to have one work order generated for all of the units instead of having 20 separate work orders to print, sign, and enter into the system as completed. On the other hand, the same school may have 4 large chillers to make the chilled water for all the unit ventilators in the school. Because a chiller failure is more critical than a single unit ventilator failure and because the repair and replacement costs for chillers is many times the same costs for a unit ventilator, each chiller would probably be included separately in your CMMS.

Deciding if something needs to be inventoried for your PM program is not an exact science. Remember that things that are included as a group, such as the fixtures in a bath room, do not need individual numbers. Equipment that is included separately will need unique identification numbers.

Step 3—Scope and Frequency

Deciding what PM tasks to do and how often to do them

By now, you should have decided exactly what pieces of equipment and building components you are going to include in your PM program. We now need to decide two things. The first is the exact procedures the PM technician will follow when performing PM. This is the "scope of work." The second is how often we will be doing each PM task, or the "frequency."

Scope—or how will you PM each piece of equipment?

Some blower motors will need 20 drops of 20W oil added to bearings, others require a high temperature grease, some utilize sealed bearings and need no lubrication at all. The PM procedures for different types of equipment will be very different. Even things that seem very similar may have very different requirements. The only right way to perform PM is do follow the equipment manufacturer's maintenance procedures.

PREVENTIVE MAINTENANCE TRUISM #4
The only right way to perform PM is to follow the equipment manufacturer's maintenance procedures.

Even similar types of equipment will have different preventive maintenance procedures. For example, two similar 1 hp circulating pumps may have different types of bearings, different coupler designs, and different materials in their rotating seals. The PM procedure that is right for one may cause damage to the other. No one knows a piece of equipment better than the engineers that designed, performance tested, manufactured, and ultimately warranty the equipment.

Get the owner's and service manuals for every piece of equipment in your facility. Fortunately, with the explosion of information available on the internet, many service manuals are now readily available on the web. Equipment manufacturers' websites often have their service manuals available for download at no cost. If you do download electronic copies of these manuals, consider printing a paper copy for your service manual library. Everyone in the maintenance shop should have easy access to these manuals.

Equipment manufacturers will sometimes provide equipment manuals with maintenance instructions by fax, or by mail free of charge. Contact all of the equipment manufacturers on your equipment data sheets and order these manuals. Having your own equipment library will not only help in setting up your PM procedures but will also be a valuable repair resource as several will include troubleshooting flowcharts and exploded parts diagrams.

Having the manufacturer's recommendations is important if you want preventive maintenance to be effective and to produce the results that are possible by doing it right. Preventive maintenance procedures can't be haphazard. Doing the right things will increase equipment life and reduce breakdowns. Doing the wrong thing may not. The wrong PM procedures can even reduce equipment life and cause equipment failures. Using the wrong viscosity of oil to lubricate bearings can wash grease out of the bearing causing them to fail. Patching a roof with a sealant that is incompatible with the roofing material can ruin the roof and void the warranty. Several studies done in the manufacturing industry have shown that up to 50% of equipment failures happened immediately after preventive maintenance was performed indicating that improper PM likely contributed to the failure.

There will be times when the manufacturer of a piece of equipment is no longer in business or when literature is not available for an older piece of equipment. In these cases you will need to rely on best maintenance practices (BMP), experience, and your own technical knowledge instead of manufacturer's instructions. Chapter 11, lists many different types of equipment and discusses in detail some general PM guidelines for each that can be used if manufacture's literature is not available. Some knowledge of lubrication theory (discussed in Chapter 5) and experience with similar pieces of equipment will also be helpful in developing procedures for those items which do not have manufacturer's instructions available.

CLAIR

In manufacturing and heavy industry, preventive maintenance mechanics use the acronym CLAIR for clean, lubricate, adjust, inspect, and repair to remember the steps to follow when performing PM to a piece of equipment. The type of lubricant, how the machine is adjusted, and other details will be different for each piece of equipment but the CLAIR principle will remain the same.

With the manufactures' maintenance instructions in hand, let's return to the equipment data sheet. You will notice the bottom of the sheet has a section titled "maintenance requirements." The PM procedures for each piece of equipment should also be added here as permanent instructions for your PM technicians to follow when performing PM. If he has any questions about what needs to be done during PM, he can open the file and refer to this sheet.

The staff performing PM should have easy access to the PM manuals. An employee break room or central shop accessible to all maintenance is a good place to locate a filing cabinet with an alphabetized set of these manuals. Many organizations with successful PM programs have also attached a copy of the preventive maintenance requirements for each piece of equipment right at the equipment. These can be typed sheets inserted in plastic pockets adhered to the side of the equipment or written on a cardboard tag attached with wire.

If you are using a CMMS to keep track of your PM schedule, the software will most likely print work orders for PM tasks that include the specific requirements for each piece of equipment. With a CMMS, the equipment data sheets and preventive maintenance records will be replaced with electronic versions. When work orders are printed for the various preventive maintenance tasks, specific PM instructions detailing which type of lubricant, filter size, or inspection procedures should be included on the work order being issued to the technician.

Frequency—or how often will you do PM to each piece of equipment?

Different equipment will need to be preventive maintenance on different schedules. Roofs are typically inspected monthly. Back-up generators are typically tested weekly but serviced annually. Circulating pumps may be greased monthly, quarterly, or even semi-annually depending on their size and hours of use.

Once again, the only dependable source of information on PM frequency is the equipment manufacturers. Equipment manufacturers will suggest how often each maintenance task needs to be done. This information should be included on your equipment data sheets for inclusion on your PM schedule.

For larger equipment under warranty, you should consider including an extensive PM inspection right before the warranty expiration date. Any problems found can be turned in for warranty repair. Large building items such as roofs, window systems, rooftop air conditioners, automated HVAC controls, chillers, cooling towers, emergency generators, or similar equipment should have a qualified service company do a complete evaluation before the end of the warranty. These dates should be included on your PM calendar, even if they are years away so they aren't forgotten.

No one can be expected to remember all of the various tasks required on every piece of equipment during PM. While the vast majority of PM tasks are very simple, there are differences between equipment which will be hard to remember. Your PM technicians will need to have the PM procedures for each piece of equipment at their fingertips. If you are using a CMMS (computerized maintenance management system) to automate PM work tickets, these procedures can be automatically included on the work tickets. Another solution is to have a preventive maintenance record binder available in the shop and require PM techs to sign this manual on the proper page each time PM is completed. My favorite solution is to post the PM procedures directly on each piece of machinery. This idea is often used by vehicle manufacturers who often include several diagrams under the hood that show lubrication points, fan belt routing, emissions system information and other instructions for anyone performing maintenance.

Figure 3-4. The front of this tag is signed each time PM is performed. PM instructions are printed on the back side.

A typed set of preventive maintenance procedures can be inserted into a protective plastic sleeve adhered to the side of the equipment or a laminated hang tag can be hung from a convenient location on the equipment. These instruction sheets or hangtags also provide a convenient place for maintenance personnel to sign and date each time PM is performed.

PM schedules based on hours of operation

Most manufacturers' maintenance recommendations will state a time interval for maintenance to be performed. The time interval could be every week, every month, or each quarter. There are some kinds of equipment, such as lawnmowers, emergency generators, and vehicles which need servicing after a certain number of hours of operation rather than after a specific amount of time. The best approach for adding these non-time-based PM tasks to your time based calendar is to estimate on what date the equipment will reach the required operating time and schedule hour meter inspections before the expected service date. By reviewing past maintenance records, it should be fairly easy to estimate how many hours of operation are clocked on each day, week, or month. The actual task of checking the hour meter reading should be placed on the calendar before maintenance is anticipated. The written work order to perform the task should be considered "open" until the work actually needs to be done. By generating a work order early, we can be sure that the equipment will be checked and that PM won't be skipped. This is one area where CMMS can really shine. Many computerized programs can keep track of how often these types of maintenance are being performed and anticipate the next required PM.

Alternatively, an "inspection" PM task can be placed on the calendar weekly or monthly to take a meter reading. Once the hour meter or odometer reaches the next PM reading, a work order should be generated to do the actual PM work. The hour readings should be written down at each inspection and will become part of your PM documentation.

As a reminder to your personnel doing the PM, placing a "Next PM Due" sticker on equipment will help. It's easy to forget if the last PM on a riding mower was at 400 hours or 450 hours. When you have your car's oil changed at a service center, they place a clear plastic sticker in the upper left corner of your windshield which tells you when your next oil change should be done. For equipment such as lawnmowers, service vehicles, construction equipment, or any equipment with a regular operator, a "next PM due" sticker or other type of tag will remind the operator to have maintenance completed.

Not all pieces of equipment come with hour meters installed. If you need to track the hours of operation on any type of equipment, it is a simple procedure to install an hour meter. Most catalogs of industrial equipment will have a variety of electrical hour meters available which can be wired into any piece of electrically operated equipment to tally hours whenever the equipment is running. Hour meters or counters can also be ordered which operate mechanically for those unusual types of equipment which do not run on electricity. When no hour meter is present, it's usually not difficult to estimate the number of hours something is operated every day and figure out when the next PM will be needed.

Choosing a CMMS

Computerized maintenance management systems (CMMS) can help to automate the work of the maintenance department. CMMS can receive maintenance requests directly from building users, generate work orders, keep track of parts inventory, schedule PM work, keep records of work completed, keep records of maintenance and repair costs, labor hours, and parts used. Because CMMS software can do so much, choosing a CMMS can be a daunting task. Unless you've worked with CMMS before, you probably don't even know what questions to ask or what features you will want to have when looking for a CMMS.

Do I really need a CMMS?

The first question to ask is if you need to computerize. If you have a good paper based system that works for you, upgrading to a CMMS might just add unnecessary complexity. Most organizations that have CMMS only use a few of the available features. Job cost tracking, parts inventory, and other features of the software are not used by many organizations that only use the software for work orders and scheduled PM.

Having a CMMS will require time each week in front of the computer printing work orders, closing completed work orders, and checking the status of outstanding work. While a CMMS can better keep track of the work that needs doing, it won't necessarily reduce the hours spent managing the work.

Setting up a CMMS will take time. Depending on how many features you plan to utilize, initial set up can take dozens of hours or more. To generate corrective maintenance or PM work orders, the CMMS will first need to know about the equipment in your buildings. That means someone will have to enter information about the equipment in your buildings. The more specific you want to be, the more time will need to be spent entering data.

Web based or installed?

Web based software is accessed over the internet using a web browser. All data are stored in the vendor's computer somewhere in cyberspace. You can access and use the software from any computer on the internet but all of the software and records are kept on the vendor's computer across the internet.

The advantage of web based solutions is that if your computer crashes, your data are safe. If the software has problems, it's not your problem. All software issues, upgrades, and maintenance are handled somewhere else by someone else and you don't have to worry about them. The disadvantage of web based software is that they can be slower to use. Even with fast internet connections, you will have to wait for data to be sent across the internet every time you do anything. The wait might only be a second or less but if you are entering 500 pieces of equipment into the computer or marking 50 work orders as complete, the seconds can add up.

The other option is to purchase software that is installed onto your office computer. The disadvantage of software installed on your computer is that you will have to do some troubleshooting whenever something doesn't work. There's also the possibility of losing your data. If you choose an installed solution, be sure to become religious about regularly backing up your files.

What does it Cost?

Web based software usually has an annual or monthly fee associated with its use. Installed software may have a one time purchase cost and an annual fee for tech support.

You should be aware that some companies make their software available as different modules. Specific modules need to be purchased depending on what you want to use the software to do. Maintenance work orders may be included in the base software but setting up a PM schedule, keeping a parts inventory, or tracking job costs may require additional modules at additional cost.

Who can enter work requests?

Some CMMS allow anyone to enter work requests via a web page. Some only allow those who have been granted access and a password to enter work requests. You will need to decide if you want every person in your facility to be able to enter a work request or if you want all work requests filtered through a supervisor or other individual. Another option is to do continue to do what you are probably doing now and accept work request via telephone and enter them into the computer yourself.

Can PM procedures be printed on work orders?

It is a nice feature for the PM technician to have the complete work procedure printed each time a work order that is generated.

Does the software have a library of PM procedures?

Having a library of generic PM templates available can be a real time saver when setting up a PM schedule. If your software already knows the generic PM tasks for most types of equipment, it can save you hours of writing procedures. Just remember that generic procedures can't take the place of the manufacturer's maintenance recommendations.

Can you enter parts used, labor hours, meter readings, etc.?

What data will the CMMS let you store and manage for use later?

Will work orders be generated automatically?

Can PM work orders be generated automatically based on a PM schedule? This is a pretty basic function of most CMMS but make sure it's included in your package.

How can work orders be delivered to technicians?

Some software only allows work orders to be printed and then hand delivered to technicians. Some will send work orders to a technician's PDA, laptop computer, or cell phone. If you have maintenance staff at several sites, can work orders be sent via fax or email? Can the automatically generated PM work orders be sent automatically to the proper location or will you have to sort through a stack of work orders each morning and decide where they should go?

Can you create your own identification number system for equipment?

Having a numbering system for equipment that makes sense to you will make using the system much easier in the future. If you can tell the type of equipment, location, and other info from the equipment ID number, a lot of time can be saved not having to cross reference ID numbers to equipment. Some CMMS generate equipment ID numbers automatically and won't let you use your own numbering system.

How long has the vendor been in business?

Stability of your CMMS vendor is very important. The last thing you want is to spend months setting up a system that you have to abandon five or ten years later because you can no longer get software support.

A short review

So far we've covered a lot of information about what needs to be done to create an effective PM program. The reason for collecting all the above data and making the abovementioned decisions is to be able to create a calendar of preventive maintenance tasks that need to be done. Let's quickly review what we've covered before we move onto creating the PM schedule:

1. We've considered our organization's goals and standards with respect to the appearance of our building and used that knowledge to decide how often we need to perform PM on the aesthetic items in our building such as offices, lobbies, guest rooms, classrooms, corridors, and on items such as carpets, wallpaper, and painting.

2. We've taken an inventory of all of the capital equipment in our facility and filled out an equipment data sheet (or entered the data in our CMMS program) for each item. We've also given each piece of equipment a unique identification number which we have placed directly on the piece of equipment.

3. We've contacted the equipment manufacturers and ordered equipment manuals for all of the equipment in our equipment inventory. We will be following the manufacturer's maintenance procedures and recommended maintenance schedule when performing PM. If we can't find manufacturer's literature for a piece of equipment, we will follow Best maintenance Practice (BMP) and experience in creating PM procedures. BMPs for several different types of equipment will be discussed later in this book.

Up until now, all we've been doing is collecting data—data about the equipment (and building components) we have in the building and data about what PM procedures and frequencies will keep that equipment in its best condition. If you've made it this far, you've put in a lot of hours. You've spent a lot of time walking your building, writing down serial numbers, marking equipment, filling out forms or typing data into your CMMS, ordering equipment manuals either online or by phone, and getting all of your data in order for the next step. The good news is all of your hard work is about to pay off.

Step 4—Making A PM Calendar

Now that you've done all the hard work, it's time to put all of the information you have collected into a useful format. Your preventive maintenance calendar will include every piece of equipment and part of your building (such as offices and rest rooms) that you have decided should be included in your PM program.

There are an enormous variety of PM calendars or PM schedules in use and there is no one right way to set up a PM calendar. All facilities are different and your PM calendars should look a little different.

Some organizations write the different PM tasks on an old fashioned paper wall calendar. So, if circulator pump #7 needs oiling every 3 months, then "oil circulator pump #7" is written on the calendar on the sixth day of March, June, September, and December. For smaller facilities, this is a perfectly fine way to set up a PM calendar. This is exactly the way I created my first PM calendar as chief engineer for a small hotel property 20 years ago.

Writing down all of the tasks to be performed can be tedious but once it's done, the same calendar can be used year after year. If you prefer the paper calendar, there are inexpensive computer programs which can make some very nice paper calendars and can automatically plug recurring events into the calendar. These are perfect for making simple paper PM calendars.

One advantage of the simple paper calendar is that it can be hung on the shop wall for everyone to see. There's no excuse for anyone to be sitting in the shop reading the paper if there's still PM to do. It's right there on the wall or shop door where it can't be missed.

Other maintenance departments use a computer spreadsheet to keep track of PM tasks and create a sort of perpetual calendar of maintenance tasks. One advantage of this method is the spreadsheet's ability to perform calculations can be used to calculate labor hours or job costs. Of course, setting up such a complex spreadsheet requires a lot of computer knowledge and experience with spreadsheet programs. Using spreadsheet software to manage PM programs can work well for small to medium organizations but has largely given way to the more user-friendly and powerful computerized maintenance management systems (CMMS) available today.

Many maintenance departments today rely on computerized maintenance management systems (CMMS) to manage their PM programs. A good CMMS software package will automatically generate work requests for routine repairs as well as generate and maintain a PM calendar. Setting up a CMMS system requires entering equipment data into the computer including the PM frequencies of different PM tasks. The computer can then generate a perpetual calendar with these data and can automatically print paper work orders on the date the work is scheduled. Preventive maintenance procedures for each piece of equipment can also be automatically included on the printed work orders.

After completing the PM tasks or other routine repairs, technicians can turn their labor hour and parts cost information into the maintenance office where a work order clerk enters this information into the CMMS. In today's wireless world, technicians can even retrieve their work orders and return labor and parts data through hand-held PDAs, laptop computers or cell phones.

CMMS systems offer another advantage to maintenance departments just starting a PM program. Because the PM work orders are generated automatically right along with the corrective maintenance work orders, they become a part of the regular weekly work stream. Introducing something new can be difficult. If a couple of additional work orders are added to the work stream and are handled exactly the way other work orders are already being handled, the introduction of the additional work is much easier to swallow.

No matter what type of calendar you choose, creating your calendar is simply a matter of taking the data you have collected and entering them, one item at a time, either into your computer or onto a paper calendar.

When putting your PM schedule to paper, try to place similar types of work, and work happening in similar locations together. Changing air filters in unit ventilators mounted above the ceiling requires getting out a box of filters and carrying a ladder. If you schedule all of the unit ventilators to be maintained at the same time, the whole box can be carried or rolled on a cart from room to room and the ladder only needs to be gotten from the storage shed once. However, if you schedule each unit ventilator on different days, it will require several trips to the shed to get the ladder and several trips to the storage room where filters are kept. Similarly, by scheduling roof top air conditioners to be inspected on the same date as the roof inspection, trips to the roof can be reduced.

In a larger organization with several buildings at different addresses, it makes the most sense to perform all PM at one location at the same time and have your PM tech work at that location for as many days as it takes to complete the PM. Having PM techs drive from building to building several times a day is simply not efficient. One of the underlying reasons to perform PM is to improve your department's efficiency.

Scheduling by the week vs. by the day

I have a small piece of advice which comes from my own personal experience. I prefer to set up PM calendars by the week instead of by the date. If a particular task is assigned for Wednesday and you have a water line to your cooling tower break underground on Wednesday, obviously no PM is going to get done. This means you have to try to catch up on

Thursday or Friday. I don't like having to catch up. I don't want a PM program to put us behind. The purpose of PM is to get ahead of the work. So I schedule all work for a particular week and I don't care what day the PM gets done during that week. We try to complete the PM in the early part of the week, so we aren't behind if an emergency happens but as long as it's completed during the week it was assigned, that's fine with me.

I've never seen a CMMS program that will let you assign work by the week. So when using a CMMS, I assign all work for Mondays and consider any work completed within five days of its assigned date to be on time. Maintenance departments are unpredictable. Scheduling by the week lets us recover when something unpredictable happens.

Other things on my PM Calendar

I also include recurring tasks on my PM Calendar which may not exactly fall into the category of PM. I include all of the required regulatory inspections that need to be completed either in-house or by outside contractors. This includes fire alarm inspections and tests, boiler inspections, elevator inspections, fire suppression system inspections, health inspections, and others. By putting these on my calendar I am reminded to call my local fire department if an inspector doesn't show up by the expiration date on the last inspection certificate. If I didn't keep all of these items on my calendar, some inspections would be missed and I would be out of compliance. Since there are often multiple agencies whose authority overlaps, a missed inspection by one agency can be cited as a violation by another.

I also include all sorts of things that come up each year that I need to remember to do. I include starting our irrigation sprinkler systems up in April, ordering parking lot de-icing chemicals in October, and writing bid specifications for landscaping contracts in January. These items show up at the proper times as work orders and are automatically printed as assigned to me instead of to one of my technicians. These aren't true PM items but I like to keep my life simple and prefer keeping one calendar instead of having to keep several.

Getting your hands dirty—The first few month of actually doing PM

Calendars start on January 1st and end on December 31st. Your calendar will probably do the same. However, the odds are 365 to 1 against your PM program actually starting on January 1st. That's fine. You can start your PM program any time, you'll be starting somewhere in the middle of your PM calendar. That's okay. Just start where you are. Don't be concerned about yesterday's PM tasks or the tasks that didn't get completed from last week. Those tasks will show up on the calendar again at the proper time. You are not behind, don't try to catch up. Just do today's work today. Slow, steady, and consistent wins the PM race.

PREVENTIVE MAINTENANCE TRUISM #5
You are not behind, don't try to catch up

When you first start performing the actual PM tasks on your calendar, it will be overwhelming. The first time performing any new task takes longer than it should and is more work than it should be. The first time through your PM program will be difficult, especially with all the work you still have to do from years of not having an effective PM program.

GETTING SUPPORT FROM UPPER MANAGEMENT

Hopefully, you have been keeping upper management informed of your work and of your intentions to develop an effective PM program for your facility. Hopefully they have been receptive and understand the benefits of such a program. Now that you have inventoried all of the facilities' equipment and put everything in a schedule, it is time to start the real work. You will need the support of upper management to make PM successful.

The first 3 months will be critical to the success of a new preventive maintenance program. The bulk of the assets in your facility will be on a weekly, monthly, or quarterly PM schedule and will be serviced at least once during these first 3 months. For a new PM program to be successful, you must be able to dedicate the staff, money, and hours necessary to keep on schedule during this crucial time. The PM program must be a priority of the maintenance department and the organization as a whole.

Upper management sets the goals and priorities for the organization and their support will be an essential part of any preventive maintenance program's success. Support from the top for your new program will ensure funding is available for the additional work which needs to be done. This support will also give the maintenance manager the authority to temporarily put other non-emergency projects on hold allowing resources to be directed toward the new program.

Businesses will support programs, if they are good for the bottom line. Management support can be gained by showing the small investment of time and money will not only pay for itself, but will also have an excellent return on investment. Hopefully, upper management was involved and supportive before you started the work of setting up the program. If you haven't already pitched your idea in detail, you should do so before you and your staff start the actual work of PM.

Your proposal to upper management should include:

1. Some brief details of what an effective preventive maintenance program will include in your building. Don't spend time on technical aspects of each task, such as how a bearing will be greased. Simply state that there are 14 circulating pumps over 1 Hp that will need to be greased quarterly. Your list of PM tasks should include all of the equipment and building components that will be included in your PM program.

2. The anticipated initial costs in materials and labor to implement a PM program. These can be estimated after you know the man hours required for each labor task. Details of different PM tasks can be found in Chapter 11.

3. The benefits of implementing a PM program. A good PM program will increase the life of capital equipment, reduce the costs of repairs, save energy, improve the experience of occupants, and reduce the number of breakdowns and other emergencies.

4. The anticipated cost savings and return on investment of a good PM program as discussed in Chapter 2.

After you have the support of upper management, be sure to keep those who have offered support up to date on the success of your program. Short weekly reports, either in person or written, on the current status of the new PM program will keep the program active in their minds and improve your chances of continued support. If you fail to remind them regularly of your program and its small baby steps toward success, it will be forgotten and other priorities will quickly take over.

PREVENTIVE MAINTENANCE TRUISM #6
If you fail to remind management of your PM program and its baby steps toward success, it will be quickly forgotten and other priorities will take over.

Keep in mind, most businesses already have a backlog of money making opportunities. There are always lots of investment opportunities available. It can be a hard sell to convince them that PM is the best investment to make right now.

DOCUMENT, DOCUMENT, DOCUMENT

The final part of your PM program will be the documentation. There are several good reasons to keep records of the PM work that you do. Let's look at a few of these reasons.

As a reference

Your records will serve as a reference in the future. When making replace vs. repair decisions or trying to diagnose an equipment failure, a good equipment history will be invaluable. A good history of equipment maintenance and repairs will be useful for budgeting and can help a technician in troubleshooting future problems. Whatever form of documentation you choose to use, make sure to include the general condition of equipment with each inspection. Although two people may judge condition differently, having a history of overall condition can provide information about how equipment is aging.

As a CYA

We've all heard the term CYA which stands for "Cover Your Ass…ets" or something like that. There will be times when a manufacturer or installing contractor will balk at making a major warranty repair by claiming the failure happened as a lack of proper maintenance. Since the manufacturer is seen as the authority, when it's your word against theirs you probably won't win. Having clear documentation of exactly what and when PM was done becomes priceless.

As a CYCoA

Similar to CYA, CYCoA stands for "Cover Your Company's Ass…ets." In this litigious society where so many people want financial compensation for everything imaginable, your PM records may protect your company from liability. There is always a risk of liability for trips and falls, indoor air quality issues, or injury due to faulty equipment. These are facility-specific things which your department has direct control over. Being able to show evidence that you have been diligent in your duty to maintain your buildings in the best possible condition can dismiss claims of negligence.

Regulatory Compliance

Your fire inspector will want to see documentation of annual fire sprinkler inspections, smoke detector testing, monthly emergency light and exit light tests, fire drills, and fire alarm tests. Your facility may be required to have annual boiler inspections, weekly or monthly emergency generator tests, annual water quality testing, annual elevator inspections, annual tests of backflow prevention devices, annual inspections of boiler smoke stack particulate matter, and many other inspections. Each department or agency that shows up on your doorstep to do an inspection will ask to see some sort of documentation of inspections and maintenance being done. A listing of the most common PM tasks and inspections required by government regulations is included in Chapter 11.

If you are using a CMMS program to manage your PM program, recordkeeping will be automated. Each time a PM work order is marked as "completed" in the computer, a permanent record of the event will be recorded electronically. Most CMMS also provide fields where you can enter things such as meter readings, parts used, labor hours, and notes. One of the biggest advantages of CMMS software is its ability to search these records and print all sorts of reports about your PM program.

If a paper work PM system is more your speed, many maintenance departments maintain their PM records in 3 ring binders divided into types of equipment. Typically a 3-ring binder will be divided into sections such as "A/C filter changes," "circulator pumps," "generator tests," "guest room PM," and "roof top ventilators." Each section would have blank PM forms that can be filled out every time a PM task is completed.

These PM forms will need to be created by you and customized for your building. For simple tasks such as changing air filters or testing the operation of emergency lights, the form can be as simple as a page of equipment numbers and locations with a space for the technician to initial and date when the work is completed. For something a bit more involved such as performing PM on a diesel powered emergency generator, the form might consist of several locations to record hour meter readings, oil pressure, coolant temperatures, fuel gauge readings, amp draw and voltages for each phase, and any other information a mechanic performing the work should check. A thorough form helps to assure that no step is missed. Some examples of PM forms are included in the appendix.

CHAPTER 3 SUMMARY

- Determining goals. The first thing we did in this chapter was establish our facilities goals with regard to aesthetic standards. Aesthetic issues such as carpet care and painting will vary more from facility to facility than mechanical issues.
- Inventory capital equipment. Using blueprints and walking through the building, we developed an inventory of the equipment that needs to be included in our PM program. The information collected has been entered on equipment data sheets or into your CMMS program. Each piece of equipment has also been given a unique identification number.
- Scope and frequency—using manufacturer's PM requirements, we have established the scope (work to be done) and frequency (how often it is to be done) for all of the equipment in our equipment inventory.
- CLAIR (clean, lubricate, adjust, inspect, repair) is an industry acronym to help the PM tech to remember the steps to performing PM properly.
- Setting up the calendar. Taking the equipment we have collected, we have created a PM schedule which includes all of the items needing preventive maintenance. Our calendar might be on paper, on a computer spreadsheet, or generated by our CMMS software. In any case, the same calendar will be used repeatedly again since all PM tasks are recurring at regular intervals. Finally we've setup a system to keep records of the preventive maintenance that we're doing.

Chapter 4

The People Who Do PM

Up to this point, we've talked about all the things we need to do to create a successful preventive maintenance program. We've identified our organization's goals, We've spent a lot of time taking equipment inventories, We've found the equipment manufacturer's maintenance requirements, and we've set up a thorough system of record keeping. If you've followed the previous chapters, you have a pretty good PM program ready to go. As you begin to use the program, you'll find some areas that will need some minor adjustments but you are most of the way there.

We've discussed the organizational tools you will need such as equipment service manuals and your preventive maintenance schedule. But when you consider all the components of an effective PM program, we haven't mentioned the most important part: your maintenance staff. Your maintenance staff will be the most important component in making a PM program a success or a failure. No matter how good your PM schedules, procedures, and documentation, a maintenance workforce that is not up to the task will make your PM program ineffective.

Maintenance personnel are often self-described "jacks of all trades—masters of none." There is some truth to the idea that no one can be an expert in everything. Unfortunately, in a disappointingly large number of maintenance shops, there is a pervasive attitude that we are not masters of any skilled trade and that less than quality workmanship is good enough. I prefer all maintenance people have the confidence and pride in their work to see themselves as "jacks of all trades" and to take pride in their ability to shift with ease from one trade to another during the day. It's true those of us who work across several trades each day will never develop a master's skill in every trade. That doesn't mean we should settle for substandard work in any trade we are asked to tackle. We should take pride in our work and take pride in the fact that few trades people ever develop our breadth of knowledge and skill across so many crafts.

I cringe every time I hear a maintenance mechanic look at their own workmanship and comment: "Can't see it from my house." or "Good enough for government work." Comments like these indicate that quality is not a priority. An environment where below average workmanship is acceptable is poison to the effectiveness of a maintenance department and to a PM program.

For a PM program to be effective, the people doing the work must have the skills and willingness to do it correctly. If I had to choose between skills and willingness, I'd choose willingness any day. I can teach skills; I can't teach attitude. Most PM tasks are basic and simple. Anyone can learn them and do them well, if they have the desire to do so. Self confidence, pride in work well done, eagerness to improve skills, and willingness to do what is needed are more important to an effective PM program than technical expertise.

PM doesn't work if it's not done well. There are many opportunities in PM to cut corners. It's easy to use whatever lubricant you happen to have on hand instead of taking the time to get the proper lubricant. It's tempting to ignore a difficult to reach bearing in the back and only grease the more accessible ones. It's convenient to skip a few daily or weekly PM tasks because the equipment was fine yesterday and should be fine today. You need to be sure the people responsible for the work take it seriously and will do their best to do it correctly. There are a few management tricks to insure PM is being done, but being able to depend on your PM staff is important.

Sometimes the problems go beyond cutting corners. It's not unheard of for technicians to sign off on PM work as completed when no actual PM was done. I can't count the times I've found dry bearings, worn belts, or clogged roof drains when all the maintenance records indicated that PM was being completed according to the PM schedule.

I've even been assured large motors were being greased monthly (after we had to replace several) and found out the custodian who claimed to be doing the greasing didn't have a working grease gun in his building.

Not all maintenance personnel are undependable. I've had the benefit of working with several who are fantastic. I became a good maintenance mechanic because I learned from some amazing mechanics. These are the people you want in your maintenance department. The guy with the broken grease gun is not.

MANAGING PEOPLE

There are a lot of people who know more about managing people than I do. Human resource (HR) professionals talk about progressive discipline, training, motivation, and lots of other stuff that's supposed to turn bad employees into good employees. I believe that you can train people to do their job better. I believe you can sometimes turn a mediocre employee into a good one by providing encouragement, challenges, and direction. I also believe truly bad attitudes can't be trained away. Repair-versus-replace decisions (see Chapter 2) sometimes have to be made about people too.

My skills are in the technical trades, construction, and in managing facilities. I am good at managing structures and equipment. As part of my duties, I also have to manage people. I think I'm a fairly good manager of people, but it's not my passion. I have a good working relationship with those who work for me. We get along well, the work gets done, and the work is usually pretty good. There's a low employment turnover rate, and I think most of the people in my department like their jobs and really don't hate coming to work. Considering maintenance is a tough field, full of unpleasant work, constant complaints (nobody ever calls to tell you something is working right), and very little appreciation, I consider myself a successful manager of people. However, I am not a management expert.

If you are having personnel problems likely to affect your new PM program, I would encourage you to get help from people who are experts. Talk to your HR department, talk to your boss, find a co-worker skilled in this area, read any of the many books on management or attend one of the many seminars available on the topic. No matter what you do, don't let a bad employee hurt your PM program. In fact, don't let a bad employee hurt your department or your new PM program.

GETTING STAFF TO BUY INTO YOUR PROGRAM

For your new PM program to be effective, all of the people performing the work of PM must consider it important. Getting them onboard should be an easy task if you can show them what's in it for them.

What's in it for them is the same thing that's in it for you. They can expect fewer emergencies, less disruptions, less difficult repairs, less occupant complaints, and a more stable, predictable work day. Instead of hiding and slinking away from your building's users, they'll be able to hold their head high, make eye contact, and feel proud of the condition of their building.

If you have a staff that cares about their work, the sell will be an easy one. A dedicated maintenance person shakes his head at equipment left to deteriorate. For those who really care, it drives us crazy to know something we are responsible for is in bad condition and we aren't doing anything about it. Maintenance people who care will prefer a system which prevents breakdowns over one which only reacts to breakdowns. We all fear new ideas, but if your staff doesn't quickly accept your new PM program, you're back to the repair versus replace decision I mentioned earlier.

THAT ATTITUDE OF CONTINUAL IMPROVEMENT

Do you remember the first truism in this book? It said that for PM to be successful, you must have an attitude of continual improvement. We all know people who are always tweaking things to try to make them better. I'm one of those people. In fact, I've got the tweaking bug so bad it might be considered a mental disease. I do so much tweaking that I have a corner of my shop filled with over-tweaked projects which didn't quite work out. My maintenance crew calls it the "corner of shame."

While I might be an example of always trying to improve things gone awry, an attitude of continual improvement is essential in the maintenance department. One great thing about an attitude of continual improvement is when a few improvements work out, it can be contagious. If you are able to make a few improvements which make life easier for your staff (and PM is a big one), they will start to embrace doing things a little differently. People fear new ideas and change. Fortunately, maintenance people tend to be tweakers. If you're not a natural tweaker, hopefully you have staff that who are. Encourage it! An attitude of continual improvement is the essence of what PM is all about.

TRAINING AND DIRECTION

If your maintenance staff is the most essential part of your PM program; and if it is imperative to embrace an attitude of continual improvement, it stands to reason that the most important area of continual improvement would be your maintenance staff.

I know training costs money and money spent on maintenance is already seen as a drain on company profits. There's also the fear that if you spend money to train an employee, the employee might leave and the training you paid for might benefit someone else. Wise people have said training someone and having them leave is still better than not training them and having them stay.

PREVENTIVE MAINTENANCE TRUISM #7
Training employees and having them leave is better than not training them and having them stay.

Even well trained PM technicians will need to know exactly which PM procedures to follow for each piece of equipment. Since each type of equipment will have slightly different PM procedures, written procedures are another important part of your PM program. In Chapter 3 we discussed getting these written procedures from the equipment manufacturers and having the procedures available in the shop, on the maintenance work order, and even affixed directly to the equipment so the technician has them at his fingertips. It's important that these procedures be easily available. Even good mechanics will rely on their instincts and experience instead of doing it the manufacturer's way if the procedures aren't right at hand. If you make them look them up the procedures in the manual each time, they never will. Written procedures need to be available where the work is being done.

Training for preventive maintenance technicians should also include instruction on how to evaluate the overall condition of equipment. The technicians doing the work will see all of your building's equipment at regular intervals and should be looking for problems and not just performing the PM tasks on each piece of equipment. PM technicians should be looking for specific problems and should let you know when an entire piece of equipment is reaching the end of its life. Their assessment along with your records of equipment maintenance and repairs will help to evaluate when it is time to make replacements.

Where to Get Staff Trained

Most PM tasks are simple and training can usually be done in house. You or more senior maintenance persons can show newer techs how some of the work should be done. For more complex pieces of equipment, manufacturers are often willing to offer PM training either at their site or yours. Spending the money to cover travel expenses, meals, and a hotel for your HVAC technician to make sure he knows how to properly maintain your

$300,000 chiller would be a wise investment. Not spending money on this kind of training is being penny wise and pound foolish.

When new equipment is being purchased, you should be able to negotiate this training as part of the purchase price. Commissioning is the term used to describe the process of turning new equipment or a new building over from the manufacturer or installing contractor to the building owner. For large equipment or systems, this process should always include training for in-house staff about the operation and maintenance of the equipment.

As technology changes, so must the skills and knowledge of your maintenance staff. There are many organizations offering training on a wide variety of topics. There is surely someone who offers training in whatever area you feel training is needed. Seminars typically last from a day to a week and can cost a few hundred to a few thousand dollars.

Training should be a line item in your annual budget. If you do not have money ear-marked for training, make sure to include it in your next budget proposal. The justification is there. Everyone in the facilities maintenance business has anecdotal evidence that a lack of training can be expensive. I'm sure you can think of dozens of times when a well meaning maintenance person created an expensive problem by trying to do something beyond their abilities. Unless your organization is willing to pay for outside experts to do the maintenance that needs to be done, they need to train your in-house staff to have the necessary expertise.

An HVAC Example

Probably the place where a lack of training is most evident is in the maintenance of many facilities' HVAC systems. Most large buildings today have some type of direct digital control (DDC) or building automation control (BAC) system to control the indoor environment. Typically these systems use complex networks of thermostats, humidistats, aqua stats, water and air flow switches, current sensors, IR occupancy sensors, and other devices to provide data to a central controller. This computer then controls every part of the heating and cooling system to maintain a comfortable environment while minimizing energy waste.

If your building is like most, your in-house staff has made adjustments over the years to solve specific problems. These control systems are so complex that most in-house technicians don't really understand the theory of operation of the system as a whole. Consequently, they don't understand how adjusting, bypassing, or switching one part of the system to manual operation affects the rest of the system.

After a few years of these adjustments being made, the system no longer works as designed and has had dozens of small band-aid modifications which have long since been forgotten. I've found flow switches jumped out, outside air dampers disconnected, variable flow controls locked wide open, variable frequency drives fixed at 60 Hz so they're no longer variable, balancing dampers adjusted to full open or full closed, and temperature sensors disabled.

Each of the dozens of people who made these adjustments was doing his best to fix a specific problem; not realizing how their fix would affect the rest of the system. Now the building is left with an HVAC system that can no longer do its job. The dozens of "fixes" that were done made things worse. And now that the system has been "fixed," occupant complaints are through the roof meaning that more adjustments will need to be made to keep occupants comfortable. Because the system is no longer operating at the designed efficiency, the gas and electric bills are also high.

A lack of training causes exactly this type of problem. If someone on the maintenance staff had been trained and understood the operation of the HVAC system, it would have been repaired properly instead of being disabled. This is an example familiar to many buildings. Maintenance technicians with good intentions will produce poor results because of inadequate training.

INVESTING IN TOOLS

Most PM tasks require the same types of tools used for regular maintenance and repairs. There are a few tools specific to PM such as thermal imaging cameras or mega-ohm meters for testing electrical insulation, but companies providing these services can usually be hired cheaper than the cost of the equipment. For most organizations, the initial investment in tools for PM will be very minimal if there is any investment needed at all.

All facilities are different in the size and skills of their in-house maintenance staff. In smaller buildings, minor maintenance may be performed by an in-house custodial staff that may not have specific trade skills. In these instances, a small investment in hand tools may need to be made. A short list would include a set of wrenches, sockets, screwdrivers, and a pair of pliers, oil cans and a grease gun. This would allow most PM tasks to be performed. Additional tools can be added as PM tasks are encountered. An oil drain pan and oil storage container for air compressor oil changes, a small thermometer to verify the discharge temperature of air conditioner units and to monitor water temperature, and a plug-in electrical receptacle tester to test electrical outlets are examples of additional tools which may be needed as work progresses.

The investment in tools cannot come without an investment in training. Expecting someone without maintenance skills to do even a relatively simple task such as changing a drive belt could result in equipment damage or even personal injury.

TRUST BUT VERIFY

When the United States and the Soviet Union first agreed to reduce our nuclear weapons arsenals, we adopted a policy of "trust but verify." We expressed to the world that we trusted our new friends, the Soviets, and had complete faith they would keep up their end of the bargain; but we were sending inspectors over just to be sure.

That's probably a good approach to take with your PM staff. There are many instances in maintenance departments when work has been signed off as completed but never actually was. There have been instances when the PM tasks which were easiest to do were completed but the hard to reach bearing all the way in the back has never been greased. The facilities manager or maintenance supervisor needs to inspect enough of the work being done to be sure it's being done and being done correctly. Even with a good dependable staff it's good policy to trust but verify. It's often said that "people do what you inspect, not what you expect."

PREVENTIVE MAINTENANCE TRUISM #8
People do what you inspect, not what you expect

Having a place on the equipment to initial when PM work is completed is a good idea. You can purchase laminated tags made specifically for this purpose. If your PM tech actually has to walk all the way across campus to sign a tag on the machine, he's more likely to do the PM task while he's there. While an initial at the equipment isn't a guarantee of work completed, it does significantly improve the chances that it will happen.

It's also important to observe your staff doing some of the PM tasks.

You probably won't need to schedule or plan for this ahead of time. As you happen to bump into your staff during the work day, make pleasant conversation, and hang out for a few minutes while they're working. It will give you a chance to see if they are doing the work properly and to do some casual training if they aren't.

OUTSOURCING PM

As mentioned before, mechanics who work across several different trades are often self proclaimed jacks-of-all-trades but masters-of-none. If your in-house staff doesn't have some of the skills needed to complete parts of your PM program, it can make sense to hire an outside contractor to perform those parts.

In the previous example of the HVAC system made useless by years of in-house repairs, the installing contractor would have been better equipped to maintain the system. The cost would have been higher to have an outside contractor make the repairs but the repair costs almost certainly would have been offset by the energy savings realized by doing it right. Hiring an outside contractor makes even more sense when you consider the value of occupant satisfaction that was thrown away by doing the repairs in-house.

Many facilities have their roofs inspected by a roofing company every year. In-house staff perform monthly visual roof inspections looking for any obvious problems, keeping debris off the roof, and making sure roof drains are free of obstructions. Once a year a qualified roofing contractor or consultant is asked to do a more thorough inspection. Roofing materials and installation methods have been changing over the past few decades and modern commercial roofs are more complex systems than most people realize. A roof can also be the single most expensive asset in a facility. Unless you happen to have someone on your staff who is unusually knowledgeable about modern roofing techniques and methods, outsourcing roof inspections almost certainly makes sense.

You might also want to include an independent third-party roofing consultant as one of your outsourced preventive maintenance people. Many roofing warranties include an annual inspection by the roofing contractor during the warranty period, which is often 10 years. While I'm not suggesting that you shouldn't trust your roofer, just keep in mind any problems that he finds will need to be repaired at his expense. He has a financial incentive to either find nothing or to recommend inexpensive band-aid repairs for any problems that are found. It's a good idea to have a roofing consultant make an inspection in addition to any inspections you might get from the installing contractor. Many roofing consultants will perform an inspection of your roof at three or five year increments and again just before the warranty expires to look for any problems that should be covered by the warranty. Independent third-party consultants can also advise you as to the proper methods your contractor should be using when making repairs.

Another reason facilities outsource some PM inspections is that many required inspections can only be completed by contractors who are licensed to do specific types of work. While you can do monthly in-house tests of your fire alarm system, only a licensed company can complete the required semi-annual inspections of smoke detectors. Your fire suppression system will need to be inspected and flow tested by a certified company each year or more often depending on your industry. Elevators, boilers, and backflow preventions are just a few more items which will most likely need to be inspected or tested by a certified company or licensed individual. There is a listing of roughly a hundred different of pieces of equipment and their inspection and testing requirements in Chapter 11.

FINAL THOUGHTS

The success of your PM program will depend more on the people doing the work then on any other factor. Maintenance budgets are often shoe string thin but you can't afford to skimp on your people. They will need the right training, the right tools, the right techniques, and the right attitude to make a PM program a success.

CHAPTER 4 SUMMARY

- The most important aspect of your PM program is your maintenance staff. Your maintenance staff's skills, training, and most importantly, attitude, can make or break a good PM program.
- If you are having difficulty in managing your PM staff, seek assistance. Your HR department, a skilled colleague, or formal training can help you to solve difficult personnel issues.

- Your PM program is important. If specific members of your maintenance staff are preventing the program's success; you may have to replace that individual. You cannot allow anyone to prevent your program or department from being successful.
- For PM staff to be prepared for their new tasks will require an investment in training. The attitude of continual improvement that makes a PM program successful also applies to continually improving the abilities of your maintenance people.

PREVENTIVE MAINTENANCE TRUISM #7
Training employees and having them leave is better than not training them and having them stay.

- While its important to have faith in the dedication of your employees, it is also necessary to find ways to verify that PM work is being completed and being competed correctly.

PREVENTIVE MAINTENANCE TRUISM #8
People do what you inspect, not what you expect

- There will be situations where it makes sense to outsource some PM tasks. Outsourcing makes sense when specialized skills are needed or when regulations and codes require certified or licensed individuals perform specific inspections and repairs.

Part II

Technical Information for Preventive Maintenance Success

The benefit of PM is that by doing a few cheap, fast, and simple things now, we can prevent expensive, time consuming, and difficult repairs later. Most PM tasks are very simple. Being a good PM mechanic does not require a degree in engineering or even years of experience working as a repair technician. Anyone with a bit of mechanical ability can quickly be taught to perform preventive maintenance. That does not mean that experience and knowledge of maintenance technology is not valuable. In fact, there are a few specific areas of maintenance knowledge that can make a significant difference in how effective your PM program can be. There are areas of PM where a little understanding can make a big difference in doing PM correctly. There are also areas of maintenance knowledge where even seasoned maintenance technicians often lack knowledge or have learned inaccurate on-the-job folklore. Hopefully, this chapter will dispel some of the maintenance myths and provide some useful technical wisdom that will make your PM team better at what they do.

Chapter 5: Lubrication Theory

One of the areas where even experienced maintenance mechanics often lack knowledge is lubrication theory. I have worked with very skilled and experienced maintenance mechanics that did not have a fundamental understanding of lubricant selection, did not know how much or how often equipment needed to be lubricated, or even how lubricants really work. Maintenance folks usually know some rules of thumb about proper lubrication. Unfortunately many of the rules of thumb amount to work site folklore and are simply wrong. Having a basic understanding of the science behind lubrication will help the building manager to make wise lubrication decisions and will make your staff better PM mechanics.

Chapter 6: Commercial Roofing

Commercial roofing is another area where there is a lack of knowledge among maintenance mechanics. While it is common for maintenance mechanics to have some working background as electricians, plumbers, or HVAC techs, few maintenance mechanics have ever worked as commercial roofers. Lack of experience with the trade combined with the wide variety of roofing materials in use makes commercial roofing an area where in-house maintenance staff are often not very skilled. Fortunately, Most organizations use outside roofers for most roof repairs so a lack of in-house roofing skills is not usually a big problem. However, if we are expecting our in-house maintenance staff to perform scheduled PM inspections, knowledge of roofing technology will make that part of our PM program more effective. When you consider that a facility's roof is often the most expensive building component in the entire facility and a roof failure can lead to so many other types of problems, it's easy to see why roofs are one of the most important areas of PM.

Chapter 7: HVAC

The third technical area in this section is Heating, Ventilation, and Air Conditioning systems. Like roofs, these systems are usually one of the most expensive systems in a commercial building. Like lubrication and roofing, HVAC is an area often misunderstood by in-house maintenance staff. Well meaning maintenance staff often understand the operation of one part of an HVAC system, but do not always understand how all the pieces fit together. In many facilities, adjustments and alterations to one part often cause performance problems in other areas.

Chapter 8: Belt Drives

Belt Drives are such a simple design and so easy for anyone with mechanical ability to understand that it may seem silly to include a whole section on the technical theory on something as simple as a belt drive. However, since drive belts account for a high percentage of equipment failures, they have earned a place here. A belt drive that is adjusted properly during PM can be expected to run dependably for a long time with few problems. An improperly adjusted drive will run hot, wear quickly, slip under load, throw belts, and fail prematurely. Since the primary application for drive belts in buildings is heating and cooling, drive belt failures are near the top of the list of failures directly effecting building occupants.

Chapter 9: Indoor Air Quality

One of the most recent and significant changes in the operation of facilities is the concept of indoor air quality (IAQ). In today's facilities, it seems that every sniffle or allergy triggers fears about a facilities indoor air quality. IAQ concerns are often brought to the building manager for investigation. Not many managers have the knowledge or tools to separate real problems from unfounded worries. Preventive maintenance practices can have more impact on a buildings indoor air quality than any other business activity and PM techs should understand how the work they do contributes or harms a building's IAQ.

Chapter 10: Paint and Protective Coatings

Few things offer greater protection to our buildings for such minimal effort and cost than a quality coat of paint. Our buildings are left out in the elements exposed to rain, salt, wind driven dirt, UV radiation, and corrosive smog and pollution. A layer of polymer only a few thousandths of an inch thick is all that stands between these damaging elements and our vulnerable facilities. Failure of this protective layer of paint cannot only allow damage but accelerate it. Recent advances in paint formulations and changes in the availability of traditional paints has been driven largely by environmental and indoor air quality (IAQ) concerns and increasing regulations on Volatile Organic Compounds (VOCs). With these changes and increase paint selection, the proper preparation, selection, and application of paint has become more complex than in the past.

Chapter 5

Lubrication Theory

TRIBOLOGY

Lubrication is one of the most frequent preventive maintenance tasks and also one of the most misunderstood. Since so much preventive maintenance work involves some sort of lubrication, it is important for PM staff to have some understanding of the science of lubrication. Most of the time, we will be able to follow a manufacturer's recommendation on frequency of oil changes, type of lubricant, and proper application methods and tools to use. However, when we do not have these recommendations, we still need to make educated choices. This requires a basic understanding of lubrication theory.

The field of engineering that deals with lubrication, friction, and wear is known as "tribology" [trahy-**bol**-*uh*-jee or tri-**bol**-*uh*-jee].

Tribologists will tell you that a lubricant has three primary functions. As expected, the first function of a lubricant is to reduce friction between interacting parts. This reduction in friction is the result of the lubricant forming a very thin film between parts which prevents the solid surfaces from actually touching.

The second function of a lubricant is to remove heat. This is obvious on systems such as a car engine or hydraulic pump where the lubricant is pumped through the system and can absorb heat and transfer the heat elsewhere. On small systems such as rotating bearings, lubricant will help to prevent the development of damaging "hot spots."

The third function of a lubricant is to suspend contaminants. By keeping dust, dirt, or metal particles in suspension, the particles are not available to scratch bearing surfaces and wear them away.

REYNOLD'S THEORY

Today's modern theory of lubrication was first proposed near the end of the nineteenth century by a mathematician named Osborn Reynolds. Reynolds was hired by the railroad industry to help in the design of axle bearings. One of his experiments showed that when a train's axel was rotating, oil would seep out of a hole drilled in the loaded side of the bearing journal. The loaded side is the side where metal to metal contact would occur. In fact, if a wooden plug was driven into the hole, the oil's pressure was great enough to push the plug out of the hole. Reynolds wanted to know how a rotating axle could produce such a high oil pressure.

What Reynolds determined was a small amount of oil was picked up by the rotating axle. This fluid was being drug into the narrow space between the axle and the journal The journal is the machinery part that surrounds and supports the axel. As this fluid was compressed into this small space at a high rate of speed, the pressure of the fluid increased.

When fluids are compressed rapidly, they develop high hydrodynamic pressures. One example is a pair of water skis skipping across a lake. If the water skis move fast enough, the fluid (water) under them is compressed so quickly that the water develops a pressure high enough to keep the skier on top of the water. Another example of hydrodynamic pressure is seen when someone does a belly flop in a pool. When the diver's body strikes the water fast enough, the water pushes back with a hydrodynamic pressure high enough to cause pain. Even a soft fluid like water can feel hard if it's compressed fast enough.

Figure 5-1. Reynolds wondered why a rotating shaft would cause oil to seep from a hole drilled in a bearing journal with enough pressure to force out a wooden plug.

The same hydrodynamic pressure that keeps a water skier above water and that causes the belly flopping diver's belly to sting was causing oil to drive the wooden plug from the hole drilled in the bearing journal.

Figure 5-2 shows a drawing of a shaft in a bearing. You will notice the shaft is not centered in the bearing but is offset to one side. This could be due to the sideways pull of a drive belt, gravity, or the shaft itself could be out of balance. All bearings experience some amount of sideways pressure and will have more clearance on one side of the bearing than on the other. The amount of bearing clearance is exaggerated in this picture. A real bearing would never have such a large gap between shaft and bearing journal.

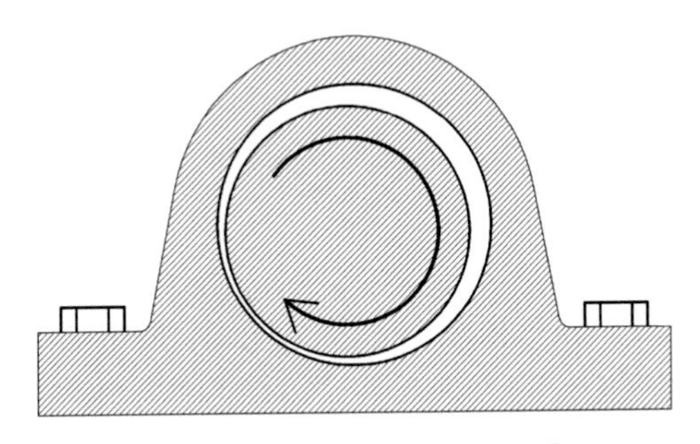

Figure 5-2. Exaggerated image of a rotating shaft in a bearing journal. Notice that the shaft is displaced to one side.

From the picture you can see if no lubricant is used, there will be surface to surface contact at the lower left side of the shaft. There is one point where the shaft will touch the bearing journal. By adding a lubricating oil to the space between shaft and journal, this contact can be prevented. This is because the pressure of the oil being dragged into this narrow space will be enough to lift the shaft away from the journal. A film of oil that had no pressure could not push against the surfaces to keep them apart. It is the pressure caused by the oil being compressed into the small space that gives the oil film the "strength" to lift the shaft away from the journal. As long as the shaft continues to rotate, the thin film of oil will maintain its

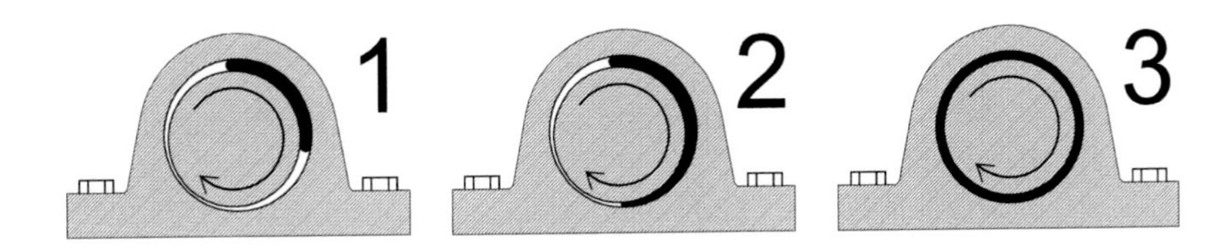

Figure 5-3. Drawing 1: Oil is pooled on one side of the shaft. 2: The rotating shaft drags oil into the narrow space between parts. 3: Compressing the oil into this space causes high hydrodynamic pressure in the oil which lifts the shaft.

high hydrodynamic pressure and will be able to keep the two surfaces apart. Typical oil pressures between bearing surfaces can be as high as 8,000psi.

The concept of a high pressure oil film is the essence of lubrication theory. The concept is that a lubricant (usually oil) is drug at a high rate of speed into a narrow restriction between interfering parts. As the fluid is compressed into this restriction, the compression creates a high pressure in the oil. The high pressure is able to lift the shaft away from the journal. The shaft then rides on a film, or very thin cushion of oil. Typical film thickness is only about .001 inches, but is enough to prevent surface contact.

Reynold's Theory Equation

Like all engineering theories, Reynold's Theory can be expressed as a mathematical equation. Let's look at a simplified version of this equation. Don't worry; we are not actually going to do any math. But just looking at the equation can give us some valuable information to help us understand choosing a lubricant.

A lubricant is supposed to keep surfaces apart. To do this the lubricant's hydrodynamic pressure needs to be high enough to lift one surface away from the other. The higher the hydrodynamic pressure, the better a lubricant can keep surfaces apart.

Even without actually doing any math, we can learn something from Reynold's equation for lubrication. Here is a simplified version of the equation that lets engineers calculate how much load an oil film can support. Remember, we want a lubricant to support the highest load possible to keep bearing surfaces from touching each other.

The load that a fluid film can support follows this equation:

$$b \times R^3 \times w \times v \div g^2$$

Where:

- b is the breadth of bearing journal (the width of the bearing)
- R is radius of bearing
- w is angular velocity (how fast the shaft is spinning)
- v is fluid viscosity (a measurement of the "thickness" of a fluid)
- g is gap width (space between shaft and journal)

We know we want a fluid film that can support the largest load possible to prevent metal to metal contact. Therefore, we want the above equation to give an answer that is as large as possible.

There are lots of things we can learn from this equation but we are only going to consider three. Let's look at each part of the above formula.

w—Angular velocity (shaft rotation speed)

Engineers use the letter "w" to represent how fast a shaft rotates. This is usually given as rotations per minute or RPM. The first thing we're going to consider is how the speed of a shaft effects lubrication. Remember, we want the highest fluid film pressure possible, that means we want to make the above equation as large as possible. Let's see how "w," angular velocity, effects the above equation. Since we are multiplying the rest of the equation by "w," we know that the bigger "w" is, the higher the lubricant pressure will be. That means that the faster a shaft rotates, the higher the lubricant pressure. A slow rotating shaft will have a hard time with lubrication. This is why slow rotating machinery usually relies on heavy greases instead of oil as the lubricant and why some slow rotating machinery have oil pumps to provide the necessary oil pressure to the bearings.

g—gap width (which is the space or clearance between bearing surfaces)

The second part of this equation we are going to consider is what happens as we change the gap width, or bearing clearance. Bearing clearance is how closely a bearing fits a shaft. We can see that the whole equation above is divided by "g," which is the gap width. That means the bigger the gap width, the lower the fluid pressure becomes. In order to have high lubricant pressure, we want to make our gap width, or bearing clearance as small as possible. Close fitting bearings will have a higher lubricant pressure than loose fitting bearings. The tighter a bearing, the better the lubrication of that bearing will be.

v—viscosity (thickness of the fluid)

The third and most important part of this equation for preventive maintenance mechanics is "v" or viscosity. Before we go any further, we need to discuss the term "viscosity." We will go into more detail on this later but viscosity is the measurement of a fluid's resistance to flow. Most people would say viscosity is how "thick," "stiff," or "sticky" a fluid is. Molasses has a very high viscosity; water has a very low viscosity.

Looking at the equation, we can see that the entire equation is multiplied by viscosity (v). That means that if we choose an oil with a high viscosity, we will get a high pressure film of oil and if we choose an oil with a low viscosity we will get a low pressure film of oil. Higher lubricant pressure is good and *higher viscosity oils will have higher lubricant pressures.*

To summarize what we have learned from this equation:

Figure 5-4. Viscous fluid on a spoon

Faster rotation
= Better lubrication

Closer bearing fit = Better lubrication

Higher oil viscosity = Better lubrication

Since we are PM technicians and we are not designing the equipment we work on, we do not have any control over rotation speed or bearing clearance. The only thing we can control is the viscosity of oil we are using. It would seem from the equation that we would always want to use the highest viscosity oil we can find because this would produce the highest fluid film pressure. Unfortunately this isn't the case.

Like many engineering formulas, there are limits to how the formula can be applied. In this case, very high viscosity oils do not work on very fast rotating shafts or on bearings with very tight clearances. High viscosity oils simply flow too slowly to be drawn into the narrow spaces in tight fitting bearings and cannot move fast enough to be drawn along by fast rotating shafts. Does that mean everything we have discussed so far was just waste of time? No.

We should understand from the equation is that at slow rotational speeds or when bearings fit loosely, we will need to use high viscosity oils. On the other hand, for fast rotating shafts and close fitting bearings, lower viscosity oil will be sufficient.

For example, many small (under 2 hp) circulating pumps and blower motors run at 1,700rpm or more, and require a light 20W oil for lubrication. Most automotive differentials have shafts that rotate below 800 RPM and require a heavy 80W or higher gear oil. Reciprocating air compressors have close fitting bearings and can be lubricated with a light 20 or 30W oil. The bearing clearance around tractor axles is usually fairly large and rotation speeds are slow. Grease with a high viscosity base oil (or even with a solid lubricant component) are usually specified in these applications.

Understanding how shaft speed, bearing clearance, and viscosity are related will help you in choosing the proper viscosity lubricant when manufacturer's specifications are not available.

CHOOSING A LUBRICANT

The way to choose a lubricant is to follow a manufacturer's recommendation. There are lots of competent maintenance people that do not understand this. For some reason, they think they know more about lubrication than the engineers who design lubrication systems for a living. They think they can do better by choosing a higher viscosity, using a synthetic where conventional oil is recommended, or by mixing oil with an after-market additive. I believe that the smartest approach is to depend on the manufacturer's recommendations all the time.

Only when a manufacturer's recommendation is not available should we make our own decisions about what lubricant we will use. Usually we will have similar types of equipment, using similar types of bearings we can use as a guide. With experience performing proper PM, you will begin to recognize which types and viscosities of oil are recommended for which types of bearings and lubricant selection will become almost automatic.

VISCOSITY

As we discussed earlier, viscosity is a lubricant's resistance to flow. Viscosity is measured by several different methods and classified by several different, overlapping systems.

Viscosity can be measured in centipoise (cP), centistokes (cSt), or Saybolt universal seconds (SUS). It can be measured at different temperatures, giving different results. To further complicate matters, oil suppliers can identify the viscosity of their lubricants in several other ways. Most industrial lubricants are sold by ISO number or by SAE weight. A comparison of several different viscosity identification systems can be found in Figure 5-5.

As we can see from Figure 5-5, an SAE 90 weight gear oil will have the same viscosity as an SAE 50 weight crankcase oil or an ISO 220 lubricating oil. Be careful when selecting oils not to confuse the different systems.

The viscosity for oil is different at different temperatures. Oil at cold temperatures will be thicker than at warm temperatures. Some oils will list a "viscosity index." This is a measurement of how dramatically the viscosity of the oil is affected by temperature. Because of the viscosity shift with temperature, many equipment manufacturers will recommend different oil viscosities for winter than they will for summer. Multi viscosity oils solve this problem because they can behave like low viscosity oils in the winter and high viscosity oils in the summer. As one example, SAE 10W-30 motor oil behaves like SAE 10 weight oils in the winter and like SAE 30 weight oil in the summer. Anytime you see a "W" in an SAE lubricating oil it indicates the oil was formulated for use in cold (winter) temperatures.

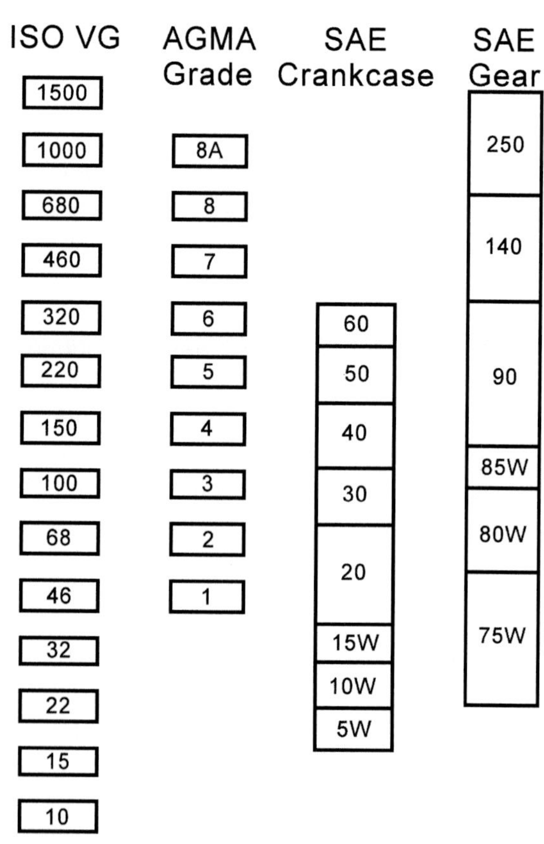

Figure 5-5. Comparison of four oil viscosity indexes

Problems with Choosing the Wrong Viscosity

If we choose an oil with a lower viscosity than recommended by the equipment manufacturer, we run the risk of having damaging surface to surface contact. As we know, oils with low viscosities may not be able to provide a high enough hydrostatic pressure to keep surfaces apart. Another problem with low viscosity oils is they tend to flow past oil seals causing leaks. In combustion engines, low viscosity oil will flow past oil rings into the combustion chamber and you will have problems associated with burning oil.

Oils with too high a viscosity will tend to leave bearing surfaces "oil starved." Thick oils cannot be pulled as quickly into the narrow spaces between bearing surfaces and will leave the surfaces dry. Rubbing surfaces will quickly be damaged. High viscosity oils can also cause excessive leaking at oil seals since the thicker oil can push against and deform soft rubber seals. Additional problems are a reduction in efficiency due to a thicker oils tendency to resist movement and problems of higher starting torque, particularly in cold weather.

Depending on the application, you may also need to consider whether your lubricant is going to be compatible with:

Rubber components
Modern low HCFC refrigerants
Other lubricants
Plastic components
High or low temperatures
Corrosive environments
Wet environments

Types of Lubricants

All of the lubricants commonly used in institutional maintenance fall into one of four categories. These categories are: 1) mineral oils, 2) synthetic oils, 3) solid lubricants, and 4) greases. All of the lubricants we use for PM will be one or a combination of these four types.

Mineral Oils

Most lubricants used in preventive maintenance will fall into the category of mineral oils. Mineral oils are produced from petroleum oil. Most of the oil familiar to us such as motor oil, penetrating oil, and most spray lubricants are mineral oils.

Mineral oils often include additives to enhance some feature of the grease. Common additives include:

- Detergents
- Dispersants
- Oxidation inhibitors
- Corrosion inhibitors
- Rust inhibitors
- Friction modifiers
- Anti-wear agents
- Extreme pressure additive
- Foam depressants
- Viscosity index (VI) improvers
- Pour point depressants

Synthetic Oils

As the name suggests, synthetic oils are man made. Synthetic oil formulations can be a combination of synthetic and mineral oil or can be 100% synthetic. Synthetic oils have been developed for high temperature applications, for longer service life, compatibility with plastic or rubber parts, or for use in unusual conditions. Synthetic lubricants used in air conditioning and refrigeration systems have been developed to mix easily with the new generation of non-ozone depleting refrigerants.

The most common synthetic lubricants are polyalpha-olefin (PAO), synthetic esters, polyalkylene glycols (PAG), phosphate esters, alkylated naphthalenes (AN), silicate esters, and ionic fluids.

The cost of synthetics can be several times the cost of mineral based lubricants. The biggest problem with synthetics is that so many skilled maintenance mechanics believe that synthetic oils will out perform the oils recommended by the equipment manufacturers. Whenever the manufacturers recommendations are available, those recommendations should be followed.

Solid Lubricants

Solid lubricants include graphite, molybdenum disulphide, Teflon®, and boron nitride. Most of these are applied as dry powders or are used as a component of greases. The most familiar solid lubricants are probably the grey graphite powder used to lubricate lock keyways and Teflon which is used in pipe thread sealant to lubricate pipe threads.

On the microscopic level, solid lubricants are made up of millions of tiny flat plates that slide easily past each other. They are used in applications where liquid lubricants would attract dirt or where lubricants need to withstand extreme pressures. Extreme pressure greases usually have solid lubricant particles mixed into the grease.

Greases

Most of the grease used today is usually composed of 85-90% lubricating oil mixed with a thickening agent, usually a metallic soap. When used as a lubricant, it is the oil component of the grease that does the lubricating. As the grease warms, it slowly releases the oil over several weeks or months providing a constant supply of lubricating oil.

Another common myth in preventive maintenance is that high temperature grease is in some way better than other types of grease. The truth is that many high temperature greases will lubricate poorly at temperatures below 250°F. High temperature greases are formulated to release their oil at high temperatures. Under normal temperatures, most high temperature grease does not get hot enough to release the oil and oil starvation of the bearings can occur.

For applications where there is extreme loading or unusually high forces applied to a bearing, specialty grease will need to be used. The fluid film of most grease cannot stand up to heavy loads and will allow metal to metal contact to occur. An example of the types of bearings subject to this heavy loading would be the pivot joints on the arm of a backhoe or the slowly rotating bearing on the dump body of a dump truck. These bearings rotate far too slowly to develop a hydrodynamic film and they support enormous forces that will cause most grease to fail.

Extreme pressure greases, designated as "EP" grease, contains tiny particles of solid lubricants such as graphite, Teflon®, molybdenum disulphide, or boron nitride. These solid particles will not ooze out under pressure and are strong enough to maintain bearing clearance even under these heavy loads.

Grease is usually used instead of oil

- For low speed bearings.
- In rolling element bearings which have large voids between rolling elements.
- Whenever a lubricant needs to "stick" to the surface, such as in open gearing.
- In wet applications where the grease tends to seal out water

Different types of thickening soaps can give grease very different characteristics. The most familiar greases are thickened with calcium, sodium, or lithium based soap.

Calcium soap based grease (also called "lime soap") is probably the most common type of grease. It is typically yellow or red in color and is often marked "all purpose."

Sodium soap based grease is another all purpose grease, typically used for slow moving equipment such as conveyors, tracks, or wheel bearings. This grease has a high drop point (temperature at which the grease no longer behaves as a solid) of 300-350°F.

Lithium soap based grease is the most common of the "high temperature" greases but can also be formulated to flow at extremely cold temperatures as low as –60°F. Lithium grease is brownish red in color and resists water well.

Barium soap based grease is similar in use to sodium based soap. It is also used for slow moving equipment and has a high drop point of about 275°F. Barium soap grease is red-yellow or green in color.

Aluminum soap based grease is a specialty grease that is extremely sticky and tends to stay in place well. It is used for surface lubrication of tracks, fifth wheels, and conveyors.

THE USEFUL LIFE OF OIL AND GREASE

Mineral oil, if uncontaminated, does not wear out. Oil that is used in internal combustion engines usually needs to be changed every 100-200 hours of operation because the combustion of fuel adds contaminants to the oil. These contaminants can cause wear to metal surfaces. Most of the oil used for lubrication of bearings is not subject to contamination. Uncontaminated oil can easily last 30 years or more under normal conditions without any reduction in effectiveness. Oil needs to be added to bearings to replace oil that has leaked or evaporated away but old oil does not normally need to removed and replaced. Some bearing assemblies have oil reservoirs that slowly supply oil to bearings through some sort of wick. The oil in these reservoirs will need to be topped off from time to time but the oil will be good for the life of the equipment.

Oil can become damaged by extremely high temperatures or contaminated in several ways. Oil contaminated with dirt and dust will damage bearings because the particles act as an abrasive. Particles can come from outside sources or from small bits of metal separated from damaged bearing surfaces. Once bearing damage starts, damage from these tiny chips cause wear to progress quickly. Water can also be a damaging oil contaminant. Experiments have shown that as little as .002% water in oil can reduce bearing life by nearly 50%. Six percent water can reduce bear-

Figure 5-6. Oil in a sealed reservoir such as the one on this circulator pump will continue to do its job for the life of the pump as long as it is not contaminated.

ing life by 83%. Water has such a low viscosity that it cannot keep bearing surfaces apart. Minuscule droplets of water in oil will allow surfaces to touch causing abrasive wear.

The only way to be sure of the condition of your oil is to have a sample taken for analysis. Usually, changing the oil is less expensive than having analysis done, which is why we change the oil in our vehicles at regular intervals, whether they really need it or not. For systems such as centrifugal chillers, changing oil is an expensive and time consuming ordeal. In this case, annual oil analysis makes sense. Oil analysis can also look for metal particles or contaminants that can give insight into the condition of chillers, generators, hydraulic systems, and other types of equipment.

HOW OFTEN DO WE NEED TO GREASE BEARINGS?

We know oil, if uncontaminated will last almost forever. This is not the case for grease. Grease is intended to continuously release its oil to the bearing and lasts only until its oil supply is depleted. This could be weeks or months and depends on temperature, size of the bearing, and how many rotations the bearing has gone through. Again the manufacturer's recommendations will tell you how often bearings will need to be greased.

In the absence of manufacturer's recommendations, it is possible to use a simple formula to figure out how often the grease in a bearing will need to be replenished. No formula can take into account all operating conditions. Bearings that operate at high temperatures, dirty environments, under heavy loads, or exposed to water will need to be greased more frequently. This formula will help to point you in the right direction. When using the formula below, "RPM" is shaft speed in rotations per minute and "diameter" is bearing inner diameter, or shaft diameter in millimeters.

$$\text{Re-greasing interval (in operating hours)} = \frac{14{,}000{,}000}{\text{RPM} \times \sqrt{\text{diameter}}} - 4 \times \text{diameter}$$

This formula works for bearings with tapered or spherical roller bearings. For cylindrical or needle bearings, multiply the answer by 5 and for ball bearings, multiply the answer by 10.

As an example:

A circulator pump motor with operating at 1725 RPM with a shaft diameter of 1 inch. We need to know the shaft diameter in mm, not inches. There are 25.4 mm in an inch so a 1" diameter shaft is 25.4mm in diameter.

$$\text{Plugging into the formula we get:} \quad \frac{14{,}000{,}000}{1725 \times \sqrt{25.4}} - 4 \times 25.4 = 1508$$

or

1508 hours, which is 62 days. If this motor runs 24 hours per day, it will need to have its bearings greased every 2 months.

Remember that this formula is for motors with spherical or tapered roller elements. If the motor has cylindrical or needle bearings you can extend the lubrication interval by five times. If it has ball bearings you can extend the re-lubrication interval by ten times.

There are other factors that can reduce grease life. Motors that are mounted vertically (shaft up or down) and motors operating in dirty environments should be re-greased twice as often. If bearings operate at temperatures over 160°F, double the greasing frequency for every 20°F over 160°F.

Bearings are usually greased using a grease gun which produces high pressure to inject grease into the bearing through a zerk or grease fitting. These fittings are designed with a ball and spring to keep contaminants out of the bearing. It is important to wipe both the grease fitting and the end of the grease gun before greasing a bearing to keep dirt from getting into the bearing.

A surprising number of maintenance techs do not know that many larger motors often have grease plugs that should be removed to allow the old depleted grease to be pushed out of the bearing as fresh grease is being pushed in. These plugs are usually opposite the grease fitting on the motor. An example of these drain plugs is shown in Figure 5-7.

You should be aware that not all motors that have threaded plugs are meant to be greased. Motor components are sometimes designed to fit several different models of finished motors. The bell end of a motor may have a grease plug that is intended to accept a zerk fitting for one motor design but not another. Inserting a grease fitting and greasing a motor with sealed bearings may just fill the electrical components with grease. Failing to add grease fittings to a bearing that needs to be greased will result in early bearing failure. If you have any questions, consult with the

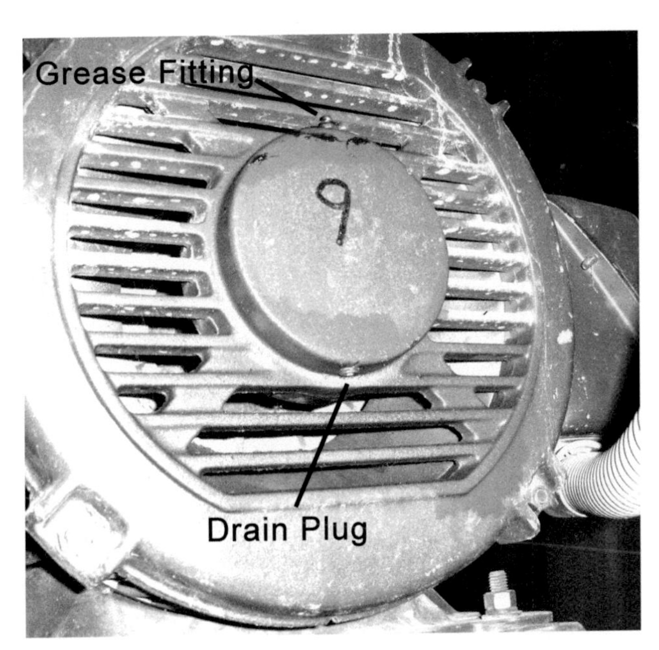

Figure 5-7. Grease drain plug on large electric motor

motor's manufacturer or your motor supplier.

One of the most common problems regarding greasing bearings is that bearings that are difficult to reach are often overlooked during PM, either intentionally or otherwise. The obvious result is that the most difficult and time consuming bearings to replace are the ones that are the most likely to fail.

TYPES OF BEARINGS

You now know more about grease and oil than most lubrication salespeople you will meet. Let's learn something about the bearings these products lubricate.

There are two main categories of bearings. The first is known as "plain" or "sleeve" bearings. The second type is "rolling element" bearings. The most familiar type of rolling element bearing is probably the ball bearing.

Plain or Sleeve Bearings

Plain bearings have no moving parts and depend on a continuous thin film of oil to keep the surfaces from making contact. Most equipment we encounter use plain bearings. Plain bearings are lubricated with oil, never with grease.

Plain bearings often have an oil hole which allows oil to enter the bearing. Oil can only travel about ½ inch to either side of the oil delivery hole. In longer bearings, oil delivery grooves are sometimes milled into the surface of the bearing to help with distribution of oil.

Figure 5-8. Plain or sleeve bearings

There are some types of plain bearings that do not require lubrication. One type is known as Oilite bearings (pronounced Oil-light). Oilite bearings are made of porous cast bronze impregnated with oil at the factory. The oil contained in the metal is slowly released to the bearing surface and is designed to last the lifetime of the equipment. Oilite bearings do not need to be re-lubricated. Many fractional horsepower motors and other types of small equipment use Oilite bearings.

Other bearings are know as "self lubricated" and are made of materials that can slide easily without needing additional lubricants. A few of the more common types are Teflon (PTFE), graphite or molybdenum disulfide (MoS2) bonded with epoxy resins, graphite, powdered metal or ceramics. Lubricating these materials with oil can actually cause damage to the bearing. This is another example of how important it is to read and follow the manufacturer's recommendations when choosing a lubricant.

Oil can be delivered to plain bearings in several different ways. Oil can be delivered through an oil hole or oil tube where it enters the bearing through an oil hole. Oil can also be added to a reservoir where fiber wicks deliver oil to the bearing through capillary action. In some machinery, fiber "wool" kept wet with oil, is packed into a "stuffing box" around a bearing and slowly delivers oil to the bearing as needed.

Rolling Element Bearings

Rolling element bearings have moving parts. Rolling element bearings consist of four parts. These are the inner race, the outer race, the rolling element (ball, roller, or needle), and a cage that holds the rolling elements in their proper place.

Rolling element bearings have a few advantages over plain bearings. Rolling element bearings can run at slower speeds than plain bearings, they can run oil starved with less damage, and they can survive contamination better since particulate matter is pushed to the side and out of the

Figure 5-9. Tapered roller rolling element bearing and bearing race

way by the rolling elements.

Rolling elements can be balls, cylindrical rollers, needle shaped rollers, or tapered rollers. Rolling element bearings are lubricated with grease, not oil. Because of the large open spaces within the bearing, grease is needed to hold the oil in place between the rolling elements. Grease is stiff enough to fill the space between the rollers and to not run out of the bearing.

Some rolling element bearings are constructed to be closed to the outside and are not intended to be greased. These bearings, called "sealed" bearings are enclosed with seals on both sides of the bearing designed to keep the factory grease inside and keep contaminants out. Sealed bearings do not have as long a life as bearings that are greased as required. Sealed bearings are often used on lower end equipment or where the equipment designers feel that bearings are unlikely to be greased due to difficult access.

Reynolds lubrication theory also applies to rolling element bearings. Under ideal conditions, each rolling element rides on a thin film of oil preventing damaging surface to surface contact from occurring. However, under extreme loading or poor lubrication, where surface to surface contact occurs, rolling element bearings are less likely than plain bearings to be damaged.

When lubricating rolling element bearings, remember that a grease gun can produce up to 10,000psi of grease pressure and bearing seals can be damaged at 500psi. You can sometimes hear a difference in the sound a bearing makes as grease is introduced into the bearing. If you want to hear this, try placing a long screwdriver on the bearing housing and the handle of the screwdriver to your ear as you pump grease into the bearing. The goal is to fill the bearing voids with grease but not to add so much grease that it is pushed out of bearing seals.

Excess friction from damaged seals can reduce bearing life by 50%. As a part of our PM philosophy of continual improvement, you might want to consider installing grease pressure release fittings or pressure release plugs instead of the factory grease plugs on some of your equipment. These devices cost only a few dollars, are available from several industrial suppliers, and will help to prevent over-greasing bearings.

Lubrication is the area of preventive maintenance that provides the most bang for the buck. The simple, quick, and inexpensive task of lubricating a bearing can save a lot of time and expense in repair costs. Lubrication is also the PM task that will show up on your PM schedule more than

Figure 5-10. Grease port extensions make it easy to grease inaccessible bearings.

any other. Having an understanding of lubrication theory will help make your PM program more successful.

Most industrial suppliers will have grease port extension kits available to relocate grease fittings to an easier-to-access location. These extension kits include threaded fittings that replace the existing grease fitting, flexible steel or plastic tubing, tube mounted grease fittings, and mounting hardware. By using these extension kits, grease fittings can be installed where they're easy to reach or even on the outside of equipment cabinets allowing grease to reach the bearings deep inside without even opening a cabinet door. Grease extension kits are an excellent part of a program of continual improvement.

CHAPTER 5 SUMMARY

- The three functions of a lubricant are to reduce friction, remove heat, and suspend contaminants.
- The modern theory of lubrication was developed by the engineer Osborn Reynolds in the late 1800s. According to Reynold's Theory, as lubricating oil is drawn into the narrow space between bearing surfaces the oil is being quickly compressed into the small space. The fast

compression of the lubricant causes extremely high hydrodynamic pressure within the oil which creates a thin, high pressure film to keep bearing surfaces from contacting each other.

- Reynold's equation tells us that at slow rotational speeds or when bearings fit loosely, we will need to use high viscosity oils. For fast rotating shafts and close fitting bearings, lower viscosity oil will be sufficient.
- Viscosity is the measurement of a fluid's resistance to flow. High viscosity oils feel sticky and thick while low viscosity oils feel thin and watery.
- There are several different systems for measuring viscosity. When selecting an oil, it is important to make sure you are using the correct system.
- The only right way to choose a lubricant is to follow the manufacturer's recommendations.
- Grease is a combination of lubricating oil and a thickener. Grease slowly releases the oil to the bearing over time. Several different formulations of grease are available for high temperature, extreme pressure, or other specialized applications.
- The most common type of rotating bearing is the plain or sleeve bearing. In this type of bearing, the shaft rotates inside a fixed sleeve which surrounds the shaft. A hydrodynamic film of lubricating oil keeps the shaft and bearing surface separated by a few thousandths of an inch while the shaft is rotating.
- The other type of rotating bearing is the rolling element bearing. The most familiar of this type are ball bearings that use rolling metal balls to separate bearing surfaces. The rolling elements can also be cylinders, tapered cones, small needles or other shapes. Rolling element bearings are lubricated with grease and not oil.

Chapter 6

Maintaining Commercial Roofs

Your building's roof is probably the most expensive single piece of equipment in your entire facility. An average roof accounts for 5-8% of a building's total construction costs. Extending the life of your roof through proper preventive maintenance can have a considerable effect on a building's long-term costs. Even medium commercial re-roofing projects can easily cost hundreds of thousands of dollars and reach into the millions of dollars. If we can properly maintain a roof and add only a few years to its life, the long-term cost savings will be significant. On the other hand, if we neglect our roof and have to make an early replacement, the negative impact will be equally significant.

The problems associated with a poorly maintained roof extend far beyond early roof replacement costs. A roof leak causes damage to the roofing deck and structure. Leaks unnecessarily tie up maintenance staff changing ceiling tiles, moving furniture and equipment, or setting up buckets to catch water. Leaks cause trip and fall accidents. Frequent roof leaks create an ambiance of a second-rate work place and affect employee moral and customer impressions. Leaks cause wet insulation which leads to a loss of insulation performance and increased energy costs. Leaks cause mold, which has recently become a major concern for employee health.

Roofing is a field which is not always well understood by building maintenance workers. The staggering variety of different roofing systems, installation methods, and materials makes roofing one of the more difficult trades for maintenance staff to master. Few maintenance workers know which repair or patch products are compatible with which types of roofing. This lack of knowledge can turn an emergency roof repair into more roof damage.

Since a roof is such a large investment and since poor PM practices can lead to such expensive repairs or replacement, it is usually cost effective to contract the services of a professional roofing consultant to make regular inspections and maintenance recommendations. During the life of the roof, professional inspections should be scheduled every three to five years and whenever a problem is suspected. To get the most of your new roof warranty, a third-party consultant should also be contracted to make an inspection just before the roof warranty expires.

During the months when a professional roof consultant is not doing inspections, it will be up to your in-house staff to inspect as part of your preventive maintenance program. These inspections will be more effective if your staff has a solid understanding of the various commercial roofing systems. The purpose of this short chapter is to make you more familiar with commercial roofing materials, installation methods, compatibility issues, and common roofing problems.

THE 4 COMMON TYPES OF COMMERCIAL ROOFS

Commercial roofing consists of an almost limitless number of materials. Fortunately there are four roofing systems that are used on more than 98% of commercial buildings. If we understand these four systems, we will be able to intelligently maintain our building's roofs on almost all of the properties we maintain. These four types include built-up roofing (BUR), modified bitumen roofing (MBR), EPDM rubber membranes and thermoplastics, and standing seam metal roofing.

BUR, MBR, and EPDM are membrane type roofs and are typically used on low slope or flat roofs. Any roof with a pitch of less than 2/12 (that is 2 inches of rise in height for every 12 inches measured horizontally) is considered to be a low slope roof. Some contractors consider anything under 3/12 pitch to be low slope. Any roof with a pitch less than one-quarter inch in twelve inches horizontally is considered a flat roof. No roof should ever be flatter than ¼ inch in 12 inches horizontally or standing water will be a problem. Even at a pitch of ¼" in 12, roof deck deflection and compression of insulation often causes low spots that collect standing water. Roofs flatter than this should have their pitch increased by installing tapered insulation board under the roof membrane when re-roofing. Tapered insulating boards are available that are tapered at a rate of 1/8 inch per foot or ¼ inch per foot. It is also a common practice to frame a new sloped roof over an existing flat roof as part of a re-roofing project.

Standing water adds weight to the roof structure causing deflection (sagging) of the roof which causes more ponding. Vegetation can become established in standing water which can grow roots into the roof membrane causing leaks. In sunlight, standing water can draw the oils out of asphalt or tar bitumens leaving a dry, cracked roof surface subject to leaking.

Built-up Roofing (BUR)

Built-up roofing systems are the oldest low slope roofing system in the country and are still the most common type of commercial roof. According to the National Roofing Contractor's Association, built-up roofs cover approximately 40% of the buildings in the United States. Built-up roofs consist of several layers of roofing felt embedded in layers of coal tar pitch or asphalt. The felt can be made of organic fibers, fiberglass, or synthetic fibers and gives the roof strength and flexibility. The layers of tar or asphalt provide the waterproof quality needed in a roof. Figure 6-1 shows the components of a built-up roof.

Tar or asphalt materials are known as bitumens. The bitumens are heated in a kettle and the material is then mopped onto to the roof deck as a hot liquid. Each successive layer of felt is rolled into the hot liquid and another coat of bitumen is mopped onto the top of each layer of felt. The result is a strong, flexible, waterproof membrane of several layers of roofing felt embedded in layers of tar or asphalt. When building conditions prevent the use of a gas fired kettle to heat the bitumen, a

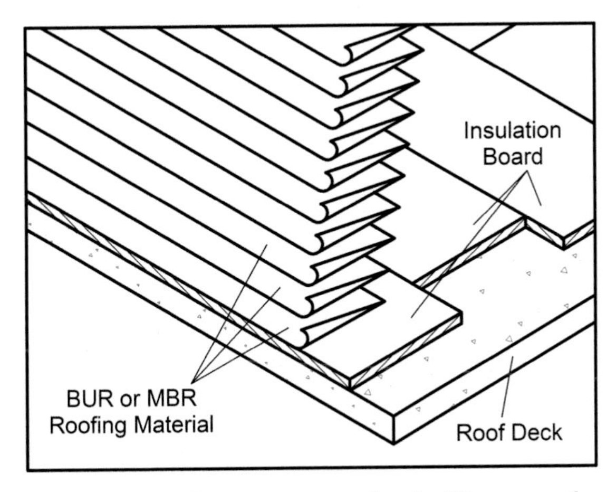

Figure 6-1. Components of a built-up roof

cold application can be done. Cold applied systems use solvents, instead of heat to liquefy the bitumen. After the roof is complete, the solvents evaporate leaving a flexible, adhesive, waterproof membrane identical to the hot mopped system.

To protect the roof from damaging UV rays that cause the roof to dry, shrink, and crack, a protective top-coat is often applied. The most common type of top coat application is to broadcast a small stone aggregate over the roof into a "flood coat" of bitumen. This layer of embedded aggregate protects the roof from UV radiation, and foot traffic.

Another common type of top coat is to embed a mineral based cap sheet into the final coat of asphalt or tar. The cap sheet is of the same material as the felt used in the other roofing layers but has a layer of fine mineral aggregate on the upper side similar to residential asphalt roofing shingles.

Built-up roofs can also have a silvery, shiny aluminum based top coat applied. This coating reflects sunlight and helps to keep roof temperatures from becoming excessively high. This type of roof coating needs to be re-applied roughly every five years.

Built-up roofs are described by the number of layers of felt used on the roof. Most BUR roofs are four layer roofs with a smaller percentage being 3 layer roofs. The number of layers has a direct effect on the life of the roof with a 4 layer roof having an expected life of about 20 years and a 3 layer roof having an expected life of about 15 years. If well maintained, a good built-up roof can last much longer. There are built-up roofs still performing their function after 40 or 50 years.

Coal tar pitch or asphalt bitumens are used for built-up roofs but these two materials cannot be used together. If coal tar comes into contact with asphalt, the tar will leach oils that will be absorbed by the asphalt. The coal tar material will harden and crack while the asphalt material will soften and liquefy. If in-house maintenance staff will be making any roof patches or sealing leaks, they will need to know which materials were used in the installation of your roof. The choice of these two products is often regional. In areas of the country with coal mining operations, tar is more readily available. For the rest of the country asphalt, a product of petroleum drilling, is more likely to be the product of choice.

Modified Bitumen Roofing (MBR)

Modified bitumen gets its name from the polymer modifiers that are added to the traditional asphalt to improve its performance. Modified bitumen roofing is very similar in appearance and in application methods to the traditional built-up roofing systems already discussed. Unlike BUR which uses layers of felt which are mopped with asphalt or tar bitumen on-site, MBR uses fabric layers which have the bitumen applied at the factory. The rolls of MBR roofing material are often called "roll roofing."

While built-up roofing is usually installed with 4 layers, MBR roofs are typically laid with only two layers. The first layer is smooth on both sides while the second layer has mineral material applied to the top surface at the factory.

The layers in a MBR system can be applied two ways: The first method, know as APP, is named for the *atactic propylene* which used as the bitumen modifier. APP is applied by what is called the "torch down" method. In this method, a torch is used to liquefy a layer of asphalt which has been factory applied to the bottom of the roofing material. The second method, SBS (*styrene butadiene styrene*), uses a cap sheet that is usually hot mopped over the smooth-surfaced first layer. The cap sheet can be smooth or can have a mineral face on the top surface. As with built-up roofs, there are also cold applied MBR systems which use solvent-based bitumens.

MBR roofing was introduced in the late 1970s and has grown in popularity ever since. Currently MBR makes up about 20% of the roofing market and that number is constantly growing. MBR has a few advantages over BUR. The addition of polymer modifiers make the finished roof layers less likely to slip due to thermal expansion or to creep under stress. It also makes the roof less brittle in cold weather. Polymer-modified membranes are tougher and more elastic that traditional built-up roofing felt.

It's often hard to tell the difference between a BUR roof and an MBR roof once the system is applied. To complicate roof identification further, there are a few hybrid systems which use one or more BUR layers under a layer or two of MBR. The expected life of a modified bitumen roof is similar to that of a built-up roof.

EPDM and Thermoplastics

About the same time that modified bitumen systems were starting to be seen on commercial roofs, another new product was being introduced. These were single-ply membrane roofs. Single-ply membrane roofs come in large sheets (usually 20′ x 100′) and are laid or glued directly to the roof deck. These large sheets are glued together at the seams using special adhesives developed specifically for single-ply roofing.

There are more than 20 different single-ply roofing materials available. The overwhelming majority of single-ply roofs installed use a material called EPDM (*ethylene propylene diene monomer*) which accounts for about 30% of commercial roofs installed today. EPDM roofs are sometimes just called "rubber roofs." The remaining single-ply materials account for less than 1% of roofs and are usually applied along with EPDM in areas where EPDM would not perform well. One common use of these other thermoplastics is around kitchen exhaust hoods. As a rubber product, EPDM can be softened by contact with grease. Therefore, it's common for PVC (polyvinyl chloride) or another thermoplastic membrane to be applied around kitchen grill top exhaust fans where cooking grease could be present.

Large sheets of EPDM are either fully adhered to the roof deck or installed without adhesive and covered with river stone ballast or concrete pavers to prevent wind uplift. Typical lifespan of EPDM roofs is similar to built-up roofing and modified bitumen and is about 20 to 30 years if maintained properly. Since this material is fairly new, many of the original applications are still in use.

EPDM can be identified by the size of the rubber sheets. Built-up roofing and modified bitumen roofs have narrow strips of roofing felt or

Figure 6-2. EPDM roof surrounded by built-up roofing

roll roofing visible. These strips will be no wider than 12 or 18" depending on the number of layer used. EPDM roofs will have sheets that are anywhere from 10 ft. to 20 ft. in width. EPDM will also be obviously more flexible and stretchy than the other two systems.

Asphalt, tar, grease, or oil will soften EPDM. EPDM should not be installed in contact with BUR, MBR, or asphalt shingles. The asphalt or tar bitumens in these products will leach into the EPDM causing it to swell, soften, and fail.

Standing Seam Metal Roofing

Standing seam metal roofs are primarily used on medium and high pitched roofs. Occasionally a standing seam roof is used in a low pitch or flat application. Standing seam metal roofing gets its name from the overlapping seam on either edge of the roofing panels that stand above the roof plane to keep the seams above any rain water that is flowing down the roof panel. A typical standing seam metal roof panel is shown Figure 6-3.

To prevent rust, standing seam panels are coated with zinc or aluminum before being painted at the factory. The expected life of a standing seam metal roof is about 25 years with very little maintenance required. In fact, there isn't much maintenance that can be done on a standing seam roof. Once the panels are secured to the deck with the hidden clips and the panels are locked together, the only maintenance that can be done is maintaining caulk at roof flashing and having the roof painted if any rust spots develop.

The maintenance free nature of standing seam roofs can be a blessing or a curse. Unlike the BUR, MBR, or EPDM, standing seam

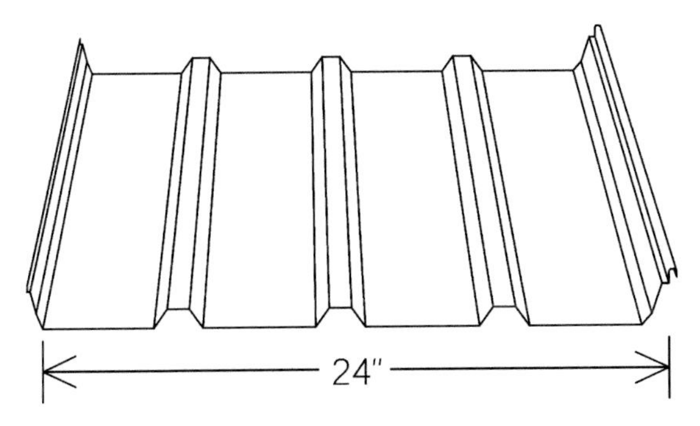

Figure 6-3. Standing seam metal roof panel

roofs don't have layers and seams that can separate and lift, there is no adhesive to dry out and crack, and the roof surface can't tear or be cut by foreign objects. The down side to standing seam roofs is that if there is a roof leak, and insulation under the roof gets wet, there is no way to cut out a section of roof, replace the insulation, and patch the roof. If insulation under a standing seem metal roof gets wet, sometimes the only way to replace the insulation is to replace the roof.

Ponding water should be prevented on any type of roof. In the rare cases where a standing seam roof is installed on a low sloped roof, it is especially important to eliminate ponding since standing water can cause the roof to rust and deteriorate.

Maintenance staff will need to be careful when walking on a standing seam metal roof. Only walk where there are purlins, or trusses supporting the panels from below. Walking in the middle of a span can dent or crimp the flexible panels creating small pools of standing water and eventual rust.

ROOF INSULATION

Almost all commercial roofs include some type of thermal insulation as part of the roofing system. BUR, MBR, and EPDM roofing membranes are often applied to a layer of insulation board which has been screwed to the roofing deck. Standing seam roofs often use fiberglass batt insulation between roofing purlins directly under the metal roof panels.

Although not as common, all of these roofing materials can also be applied directly to the roof deck and the insulation can be installed inside the building, under the roof structure. In a few extremely cold areas, insulation board is sometimes even laid on top of the roofing material and, held in place with stone ballast. This inverted roof (IR) system keeps the roof membrane on the warm side of the insulation so that the membrane isn't damaged by thermal contraction when the roof is brittle most likely to crack from the cold weather. The insulation board in an IR roof also serves to protect the roof membrane from UV light and physical damage.

There are more types of roof insulation in use on commercial buildings than types of roofing. It's not important for the PM mechanic to know all of the types but I've included a quick description of the most common types here for the curious. It is important for the PM mechanic to understand how roofing insulation can fail.

The first and most common problem with roof insulation is when there is a roof leak and insulation gets wet. Some of the closed cell insulation boards can survive being temporarily wet with little impact on performance or condition of the roof. Other types absorb water readily and loose their insulating value or will compress creating low spots and ponding. Areas of ponding water are going to be right at the same location as the leak which caused the insulation to get wet in the first place. The extra weight of saturated insulation can also cause deflection of the roof structure which will also increase ponding right at the leak location.

There is no accepted method to dry insulation trapped under a roof membrane. Once water damage occurs, the only solution is to cut out that section of roofing and install new insulation. For most of us, this is something we would not attempt to handle in-house and an outside contractor would be hired to make these repairs. In many instances wet insulation goes unnoticed and the wet insulation is left in place for years. When insulation is left wet, the insulating properties of the material will be lost and wood or steel roof decks can rust or rot away unnoticed. Wet insulation, especially insulation made of organic materials, is also an excellent food source for mold.

Some types of insulation can also be damaged by foot traffic. The fibers of a few insulation boards will separate and compress under load. I once witnessed a very small EPDM roofing job that had to be redone because the foot traffic of installing the roof destroyed the insulation board. This was a very small roof with skylights, AC units, and other penetrations in a very tight space. Two roofers had very little place to walk and spent all day shimmying around all of the roof's equipment cutting and installing small panels cut to fit each space. When they were done, the insulation board had been completely destroyed and the roof had to be redone, this time with a different type of insulation board.

It would seem that this would not be a problem if roofers would choose one of the insulation boards that resists moisture and stands up well to foot traffic. The problem is one of fire ratings and building codes. Most of the closed cell materials that resist water can add fuel to a fire and must be installed with some sort of fire proof underlayment to separate the roof from the structure. Gypsum board, a material very similar to drywall, is the most commonly used fire proofing material. Gypsum absorbs water and will fall apart when wet. So if we choose an insulation board that resists moisture well, we end up with a fire barrier that won't. There are mineral based insulating boards that are water resistant, durable under foot traffic, and fire proof. However, these boards are so rigid that they can't conform to any drainage contouring of the roof deck and many roof structures are not designed to handle the weight of these boards. Mineral insulation board is also more expensive than other types. There is no perfect roofing insulation that solves all of the potential problems.

Common Types of Roof Insulation

Polyisocyanurate foam board (a.k.a. PIR or "poly-iso")—a rigid closed cell foam board often manufactured with a foil face to help with radiant heat loss. Although color can vary, this is usually seen as a soft, pale yellow foam board sandwiched between two layers of silver colored foil.

Polyurethane Foam Board—a closed cell foam board very similar to poly-iso in appearance and use. Requires a layer of gypsum to be applied under the board to achieve proper fire ratings.

Polystyrene Molded Expanded Foam (a.k.a. MEPS, "bead board")—a closed cell foam board called "bead board" because the board is made of tiny white polystyrene beads pressed together. Although each closed cell "bead" is water resistant, water can enter the spaces between beads and saturate the insulation board. Most often used with stone ballasted single-ply membrane systems. Can be dissolved by tar or asphalt bitumens. Available with a variety of facings. Requires a layer of gypsum board for fire rating.

Extruded/Expanded Polystyrene Board (a.k.a. XEPS)—similar to MEPS but slightly more dense.

Cellular Glass Board—a very ridged insulation board made of a closed cell glass matrix with asphalt coated Kraft paper facing. Exceptionally high compressive strength takes foot traffic well. It is nonflammable and requires no additional fire protection. Boards flex very little and do not conform to irregular roof decks. Kraft paper facing absorbs moisture. Has the appearance of grey stone.

Mineral Board—also called perlite board. Manufactured of expanded perlite ore, cellulose fibers, asphalt and starch binders. Absorbs and holds moisture.

Wood Fiberboard—Used in combination with the various foam board materials to make a composite material that has good insulating properties and high rigidity and compressive strength. Wood fiberboard was commonly used alone as roof insulation board up until the 1960s.

Glass Fiberboard—The more modern version of wood fiberboard. Used as a composite material with other foam insulating boards and occasionally used on BUR re-roofing applications where the compressible fiberboard will conform to a rough aggregate coated roof. Readily absorbs water.

Fiberglass Batt Insulation—soft flexible blankets made of glass fibers, often with Kraft or foil facing and sometimes encased entirely in plastic sheeting. Used in the space of building structures under roof decks or between purlins when installing standing seam metal roofing. Excellent thermal properties, absorbs and traps water readily.

FASTENERS, FLASHING, AND ROOF PENETRATIONS

Since our interest is preventive maintenance, our interest in roofing is how to best maintain them. We will talk about the specifics of roof inspections and preventive maintenance tasks in the next section of this chapter. By now, we should have an understanding of what roofs are made of, how they are installed, and a few of the problems all of these roofing systems can have. There are more components of roofing systems which we need to be familiar with if we are going to be effective in maintaining roofs. These are fasteners and flashing.

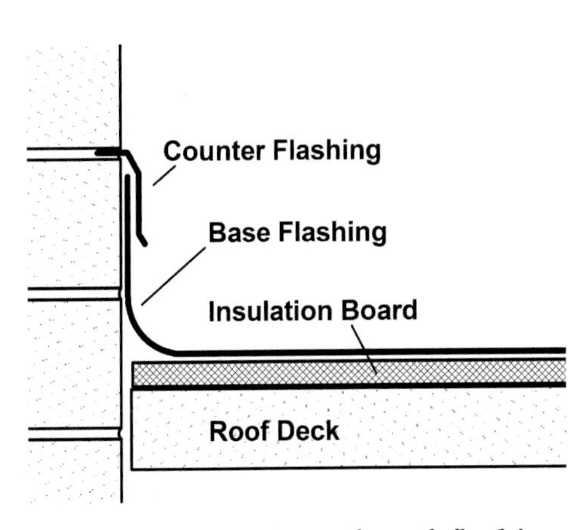

Figure 6-4. Drawing of roof flashing detail

Roof fasteners are used to secure roofing materials to the roof deck. Flashing is the rain-proof transitions between horizontal and vertical surfaces, between the roof and roof penetrations, and around roof mounted equipment.

Roof fasteners and flashings contribute to roof problems more than the roofing materials themselves. Old roofs which have outlived their usefulness will develop leaks as the roofing materials crack, separate, and split. Roofs that are still viable rarely leak in the field of the roof. Most leaks on roofs that are still within their useful life will be caused by faulty fasteners or flashing.

Fasteners

As we now know, membrane roofing systems (BUR, MBR, EPDM) are typically applied on a layer of thermal insulation board which is installed onto the roofing deck of the building. Roofing fasteners are used to secure this insulation board to the deck surface. When the roofing materials are applied directly to the decking, the fasteners can be eliminated from the system. Since roof decks can be concrete, wood, or steel, a variety of fasteners are used to anchor the insulation board to the deck.

Screws, toggle bolts, or steel pins are driven through the insulation board and into the roof deck, depending on the roof deck material. The fasteners typically use large flat washers under the head of the fastener to hold the insulation board down. When the fasteners are driven, they are dimpled slightly into the face of the insulation to prevent the fasteners from standing proud of the insulation creating a sharp point which could damage the roof membrane.

The problem with roof fasteners is that roof fasteners sticking out higher than the insulation board can puncture the roof membrane from below. If a fastener breaks off under the roof, it is likely to cause a hole and a leak. If fasteners are breaking off or pulling out of the roof deck, it is often because they are being corroded by water saturated insulation. Fasteners can also protrude above the insulation when insulation board has been compressed from foot traffic or water saturation. One of the important things to look for during a roof inspection are the tell tale bumps of roof fasteners pushing up under the roofing material.

Flashing

The purpose of roof flashing is to transition from one part of the roof system to another. While there are countless varieties of roof flashings, they all work in the same way. They all include a base flashing attached directly to the roof membrane which continues from the horizontal surface up to a vertical surface such as a skylight curb, parapet wall, or roof penetration. The base flashing is usually made of the same material as the roof. The second component of the flashing system is the counter flashing which attaches to the vertical surface and overlaps the base flashing. Counter flashing is usually formed aluminum sheet but can be stainless steel, galvanized steel, copper, or even lead. By overlapping the counter flashing over the base flashing, any rainwater running down the vertical surface will be directed on top of the roof membrane.

When inspecting a roof, the flashing is a critical component that needs to be inspected. Some common problems found when inspecting flashings are:

- Parapet wall cap flashings blown off by wind
- Parapet wall cap flashings with failing caulk at overlapping joints
- Mortar missing where counter flashing is set into a brick or block mortar joint.
- Failing caulk at joints between pieces of counter flashing
- Base flashing that is no longer tucked up under the counter flashing
- Counter flashing bent by the wind
- Base flashing with holes caused by wind-blown objects. Since base flashing isn't always supported from below, it tends to puncture easily.
- Base flashing fasteners falling out of the vertical surface (most common with EPDM)

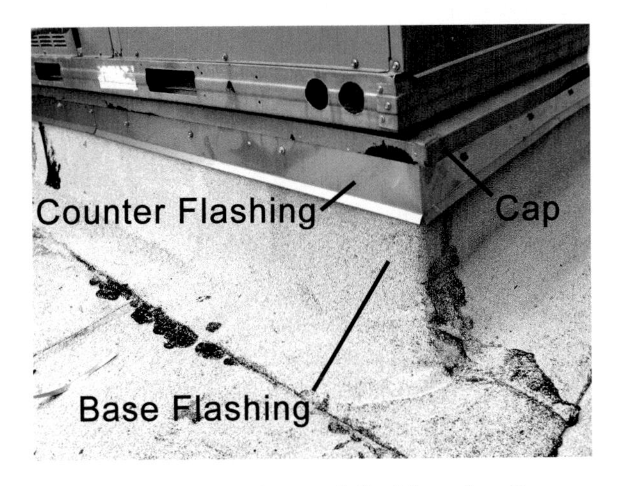

Figure 6-5. Photo of flashing detail

ROOF WARRANTIES

Roof warranties typically come in three types; materials only, materials and labor, and "full system" warranties that cover the entire roofing assembly. Materials only warranties aren't worth very much since most roofing problems are caused by poor installation, not poor materials. Most labor and material warranties are pro-rated. This means the dollar amount available for repairs decreases each year as you get nearer to the end of the warranty. These prorated labor and material warranties provide more coverage than materials only warranties but offer very little warranty protection when roof failure is most likely, near the end of the warranty period.

There are also no dollar limit warranties (NDL) available for most manufacturers' roofing products in 15 or 20 year terms. NDL warranties need to be purchased at an additional cost to the roofing project. In the event of a roof failure, NDL warranties will cover the entire cost of roof repair or replacement until the day the warranty expires. In many cases, if you purchase a 20 year NDL warranty, the manufacturer will specify an extra layer of roofing material be installed. NDL warranties cost more than traditional roof warranties and are only purchased for a small percentage of roofing projects but may provide the best value.

Roof warranties are typically offered by the manufacture of the roofing materials. The installing contractor often offers a 1 or 2 year labor warranty in addition to the manufacturer's warranty. After the initial contractor's warranty period is over, the manufacturer's warranty takes over.

Most roof warranties have an extensive list of conditions that must be met by the building owner or the warranty can be voided.

A list of common warranty exclusions and conditions follows:

1. "The building owner must notify the manufacturer or supplier, in writing, within a [specified time period] of any roof leak." Most facility managers call their roofing contractor and request a repair over the phone. Failing to notify the manufacturer or supplier in writing can leave you without a warranty if repairs become extensive and the contractor decides to walk away.
2. The manufacturer or supplier shall not be responsible for incidental or consequential damages." This protects the manufacturer from responsibility for ceiling tiles, carpet, or equipment damaged

by a leaking roof and can even extend to wet roofing insulation board that was installed at the time the roof was installed.

3. "All judgments, decisions, and determinations are those of the manufacturer or supplier only."
4. "The owner must use reasonable care in maintaining the roof." According to number three, it seems the manufacturer gets to decide what constitutes "reasonable care."
5. "...not responsible for roof components not supplied by the manufacturer or supplier." Including flashing, insulation, fasteners, etc. When any components are used that were not provided by the roofing manufacturer, the warranty may be void.
6. Many warranties exclude damage due to wind of "gale" strength or more. The facility manager needs to realize that The Beaufort Wind Scale lists a moderate gale as 32-38 mph.
7. Most warranties prorate the dollar limit over the life of the warranty, not the expected life of the roof. If you need a new roof after 5 years with a 10-year warranty, you will only be able to recover 50% of the cost of the roof (often only 50% of material costs), even though the roof has failed at only 25% or less of its expected life.
8. Warranties are often not assignable to any new owner unless the manufacturer approves, a fee is paid, and all repairs recommended by the manufacturer are completed by the owner.
9. Roof inspections must be completed on a regular schedule specified by the manufacturer. Failure to complete (and document) inspections can void the warranty. Other warranties require that the owner take specific steps such as recoating the roof every 5 years. The warranty could be voided if this wasn't done, whether the condition of the roof justified the work or not.
10. Many roof manufacturers will not warrant their roofs if they are installed over an old roof.

Typically, if only small occasional repairs are needed during the warranty period, your installing contractor will not hold your feet to the fire concerning all of the warranty conditions. However, if a roofing installation needs extensive repairs or replacement, it is not unusual to end up in litigation. It is important that all warranty conditions be considered when creating the roofing part of your PM schedule. If you find yourself needing to enforce a roof warranty, your preventive maintenance records can be very important to document compliance with each specific condition.

PREVENTIVE MAINTENANCE OF THE ROOF

The most important thing to remember about maintaining a roof is: If the roof is under warranty, don't touch it. To maintain your warranty, have the installing contractor make all repairs during the warranty period. Even when the warranty period expires, you will still likely want to let a professional roofing contractor make most of the repairs. A roof is simply too expensive and mistakes too costly for a do-it-yourself approach unless you happen to have commercial roofing expertise in-house.

Minor repairs, such as caulking flashing, reattaching loose flashing, or re-coating a roof can usually be handled in-house, as long as you are confident that you know which mastics, adhesives, and sealers are compatible with your particular roof. This is information you should get from the installing contractor whenever you have a new roof installed.

The backbone of the preventive maintenance program for commer-

Figure 6-6. This broken drain strainer is letting acorns into the storm drains where they may cause a clog. One of the many items to checked during a PM inspection.

cial roofs is the roof inspection. All roof PM starts with an inspection to finds small problems which need to be corrected. Inspections should be done by roofing professionals every three to five years. Between these professional inspections, your in-house maintenance staff should be doing monthly inspections to look for any current or future defects, to remove debris that may cause tears or punctures, and to keep roof drains, gutters, and scuppers free of trash. By inspecting each month, small problems can be found and repairs made while they're still small and inexpensive.

Having a thorough roof inspection completed regularly not only allows you to make minor repairs while they are still minor, it also gives you an estimate of the remaining roof life. A roof replacement is often the most expensive type of maintenance project for a facility and requires budgeting several years in advance. Knowing that a major roofing project is on the horizon allows the necessary advanced planning to complete the project when it is needed instead of years after the roof has started to fail.

The following roofing inspection checklist can be used by your in-house staff during their regular roof inspections. Now that we have a better understanding of roofing systems, most of these items on this checklist should be self explanatory.

Figure 6-7. Separating caulk on this coping joint will let water into the wall.

Figure 6-8. Any stones that have gotten past the gravel stops must be removed from the drain.

Built-up roofing (BUR) & modified bitumen roofing (MBR) and single-ply (EPDM) PM inspection checklist

- ❑ Check membrane material for any loss of surfacing (mineral surfaced)
- ❑ Check stone ballast for movement due to water or wind and redistribute if needed
- ❑ Check all seams for separation or softening
- ❑ Visually inspect roof surface for appearance of insulation board fasteners
- ❑ Walk entire roof and "feel" for any collapsed, soft, or apparently wet insulation.
- ❑ Visually inspect roof surface for punctures
- ❑ Visually inspect roof for any cracks, tears, or bubbles.
- ❑ Check the condition of any roofing tar, asphalt, or other sealants used to patch leaks
- ❑ Check for any wrinkled areas of roof indicating that the roof membrane is shifting (usually downhill)
- ❑ Check for material shrinkage resulting in end laps that no longer overlap

- ❑ Check for membrane brittleness or drying
- ❑ Check condition of coating (if applied)

Flashing, fasteners, penetrations PM inspection checklist

- ❑ Check all counter flashings for missing mortar in mortar joints
- ❑ Check felt or MBR base flashing for holes at all flashing locations
- ❑ Check felt or MBR base flashing fasteners at vertical surfaces (if used)
- ❑ Check all base and counter flashing to be sure it is adequately anchored in place
- ❑ Check end lap joints of counter flashing or parapet cap for sealant between pieces

Standing seam metal roofing PM inspection checklist

- ❑ Inspect all panels for holes or penetrations
- ❑ Check all standing seams for separation of panels
- ❑ Check all panels for signs of rust (have repainted before rust spreads)

All roofing types PM inspection checklist

- ❑ Are any roof drains clogged with balls, debris, gravel dams, etc?
- ❑ Are gutters or leaders clogged or filled with debris?
- ❑ Check all penetration pitch pockets for adequate potting material
- ❑ Check surface of roof for sharp objects or other debris
- ❑ Check roof for standing water or ponding
- ❑ Check flexible expansion joints for splits, cracks, or drying
- ❑ Visually inspect the condition of flashings at all roof penetrations
- ❑ Check scuppers for open seams, cracks, and to be sure they are free of clogs

There is one roof defect that is significant and common enough that it deserves to be mentioned alone. This is the problem of water saturated insulation trapped under the roof membrane. Identifying wet insulation, where it can't be seen, can be difficult. Sunken areas of the roof can give clues, so can a "spongy" feeling as you walk across the roof. But neither of these is a sure way of identifying all sub-membrane water.

The traditional method of looking for saturated insulation is to take core samples. A core sample is a plug of roofing material punched

Figure 6-9. Pitch pockets need to have tar or asphalt pitch added as the material shrinks.

through all the roof layers. A grid of dozens of core samples must be taken to look for the presence of water under a large roof. Core samples leave holes that must be patched that could leak later causing more roof problems.

Fortunately, technology has provided other methods of finding water under the roof. The most common method used by roofing consultants, is thermal imaging. Water trapped under the roof will tend to warm and cool slower than the rest of the roof. By looking for hot or cold spots on the roof, it is possible to see the exact location of water under the membrane. Thermal imaging should be part of your roof inspection program and should definitely be used near the end of your roof warranty to get the maximum value from your warranty.

CHAPTER 6 SUMMARY

- A roof is the single most expensive building component on most buildings. Extending roof life by proper preventive maintenance can have a significant impact on long-term building operating costs.

- There are 4 common types of commercial roofs. These are built-up roofs (BUR), modified bitumen roofs (MBR), Single ply membranes (the most common being EPDM), and standing seam metal roofs.
- Built-up roofing is typically used on flat or low sloped roof applications. BUR roofs are made up of several layers of roofing felt embedded in layers of liquid asphalt or tar.
- Modified bitumen roofs are very similar to BUR roofs. MBR roofs have the bitumen (asphalt or tar) impregnated into the roofing felt at the factory and use asphalt or tar with modifying additives to increase flexibility and improve roof performance.
- EPDM rubber roofs are applied as a single sheet over roofing decks.
- Standing seam metal roofs are made up of interlocking steel panels and are most suitable for medium an high sloped applications.
- There are two types of bitumen used in BUR and MBR roofs. These are Asphalt and Tar. These two materials cannot be used together. If asphalt and tar are used together, the asphalt will absorb oils from the tar. As a result, the tar will become dry and brittle while the asphalt will become swollen and soft.
- Roof deck insulation can be ruined by water penetration. Water will reduce the insulating properties and can also cause collapse of the insulation board.
- A common cause of roof problems is insulation to roof deck fasteners poking holes in roofing materials from below. This can happen if fasteners move or if insulation board compresses causing fasteners to stand proud of the surface of the insulation board.
- Roof flashings, counter flashings, pitch pockets, copings, and other transition materials are a common cause of leaks and should be thoroughly inspected.
- Roof warranties are available in several different types. Most roof warranties are pro rated for the duration of the warranty and have many maintenance conditions that must be met for the warranty to remain in effect. Therefore, it is very important to consider the warranty conditions when creating a PM program for a roof and to keep excellent documentation of maintenance performed.
- All PM programs should include regular (every 3 to 5 years) in-

spections by an independent roofing consultant. This inspection should include looking for hidden moisture in roofing insulation. This inspection is most commonly done using thermal imaging cameras.

- An independent roofing consultant should be contracted to perform a roof inspection near the end of a roofing warranty so that any problems can be solved by the installing contractor before the warranty expires.

Chapter 7

HVAC Systems

Air conditioning is one of the areas where proper preventive maintenance can have one of the biggest impacts on a facility's long-term costs. On average, a facilities HVAC system accounts for about 11% of total construction costs. Proper PM of HVAC equipment is important, but these systems are often complex and rely on engineering principles that not all maintenance technicians understand. If you are fortunate to have an HVAC technician on your staff, PM of your air conditioning equipment will be a simple in-house task. If you are relying on maintenance technicians from other trades your pm program will be more successful if they have a fundamental understanding of air conditioning equipment and its operation.

Federal law requires that anyone who does work on the closed refrigerant circuit of air conditioning systems be certified by the EPA in the proper procedures for working with refrigerants. Since most refrigerants contain chemicals that can contribute the depletion of the ozone layer, and global warming, there are specific procedures that must be followed to prevent the release of refrigerant to the environment. Most PM can be performed without doing any work on the closed refrigerant circuit, but any work done on the closed system will need to be done by someone with the proper EPA certification for the type of refrigeration equipment being serviced.

As part of striving for continual improvement in your PM program, consider offering HVAC training to your maintenance staff as part of investing in your people. Many community colleges and local vocational schools offer evening courses in air-conditioning installation and service.

When the necessary skills are not available in your department, it is usually money well spent to sign a PM maintenance contract with a service company who does have the required skills. A service company will cost more than doing PM in-house, however the savings realized by doing PM correctly will more than offset the cost of the PM contract. In one facility where I worked, signing a $8,000 PM contract on food service equipment reduced the cost of repairs from $23,000 to $7000 in the first year.

This chapter on air conditioner operation is not to be a substitute for formal training in air conditioning service. Sometimes budgets or other factors require that we start our PM program with in-house staff, even if they may not have all the skills we would like them to have for the job. Remember, some PM is better than no PM. As your PM program becomes successful, you will be able to hire qualified HVAC service staff, contract the work out, or get your existing staff the proper training to perform the work. In the mean time, this chapter's overview of HVAC systems will help to explain some of the preventive maintenance work required.

Air conditioning and refrigeration systems are sized according to the amount of heat they can remove from a building. The two commonly used measurements of cooling or heating capacity are Btus per hour (British thermal units) or tons. One ton of cooling is equal to 12,000 Btu/hr, which is the amount of cooling required to convert one ton of water at 32°F to ice in a 24-hour period. Window or thru-the-wall air conditioning units are usually 2 tons or less, rooftop units (RTUs) are typically between 5 and 30 tons of cooling and centralized chillers are usually sized between 100 and 500 tons. Of course these numbers can and do vary.

REFRIGERATION MACHINERY

Most of the air conditioning systems in use in facilities use the same technology that was invented by the engineer Willis Haviland Carrier in 1902. The refrigeration machine was invented to control temperature and humidity during printing operations and was used exclusively for industrial operations for the next two decades. Air conditioning wasn't used for human comfort until Carrier air-conditioners were installed in the J.L. Hudson Department Store in Detroit in 1924. Air conditioning caught on and most of us cannot imagine a world without it today.

Small modifications have been made over the years but the basic design and operating principals have remained the same. Before we get into the preventive maintenance of HVAC systems, let's go over a short explanation of how a refrigeration circuit actually works. Please refer to Figure 7-2.

Figure 7-1. This 20-ton rooftop air conditioning unit has enough cooling capacity to produce 20 tons of ice in 24 hours.

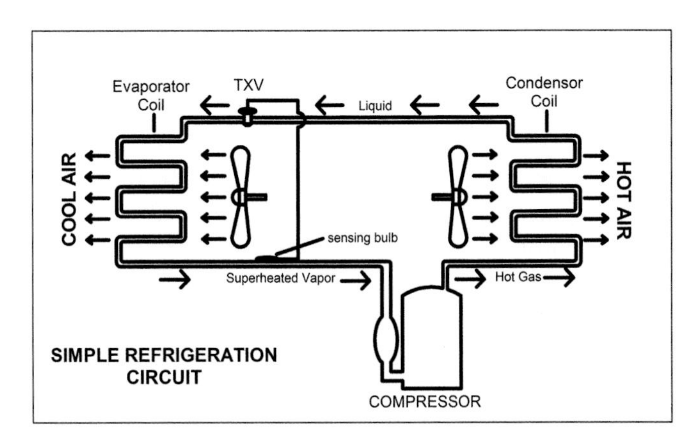

Figure 7-2. Components of a simple refrigeration circuit

The refrigeration system is a closed circuit. Nothing gets in and nothing gets out. The system contains refrigerant, lubricating oil for the compressor, and nothing else. Filters and dryers are usually included to make sure no other contaminants or water are present in the closed system.

The refrigerant is moved through the refrigeration circuit by the compressor. The compressor draws in refrigerant vapor on its inlet side (called suction or low pressure side) and compresses it to a high pressure which is discharged on the outlet side (discharge or high pressure side) of the compressor.

This high pressure refrigerant vapor continues through the refrigeration circuit until it reaches the condenser coil. Air is blown across the condenser coil to remove much of the heat from the refrigerant thereby cooling the refrigerant inside the coil. As the refrigerant cools, it condenses into a liquid. This is very similar to water vapor condensing on the outside of a cold glass on a humid day. In practice, the condenser coil is located outside of the building or in some way designed to expel the heat to the outside environment.

The cool liquid refrigerant, still at high pressure, is moved along from the condenser coil until it reaches the expansion device. An expansion device is nothing more than a small restriction that only allows a small amount of refrigerant to pass. There are different types of expansion devices. The simplest is called an orifice and is just a round disk with a small hole in the center. Different size holes are available for different size air conditioning equipment. The most common type of expansion device on commercial equipment is the thermostatic expansion valve or TXV. A TXV has a temperature sensing bulb and internal bellows and is able to adjust the size of its orifice to maintain a constant refrigerant temperature in the evaporator coil. Electronic expansion valves (EXVs) are an electronically adjustable version of an TXV.

Since the expansion device only has a very small opening, the refrigerant sprays out of the device in a mist of small droplets comparable to the spray from an aerosol paint can. The spray of refrigerant is now at low pressure and is moved along into the evaporator coil, sometimes referred to as a direct expansion (DX) coil. The evaporator coil is the coil where cooling takes place. Liquid refrigerant, at low pressure evaporates in the evaporator coil and becomes a low pressure vapor. As the refrigerant evaporates, it absorbs heat from the air passing over the coil. When the air gives up its heat to the expanding refrigerant, the air is cooled. This is the fundamental principal that makes air conditioners work: *as the refrigerant evaporates, it absorbs heat from the air.*

If you want to see an example of an evaporating liquid absorbing heat, put a little rubbing alcohol on the palm of your hand. When the liquid rubbing alcohol evaporates it makes your hand feel cold. This is because the evaporating alcohol absorbs the heat from your palm. All liquids do this. Some do it better than others which is the reason some liquids make good refrigerants and some don't.

Leaving the evaporator coil, we have a warmer refrigerant vapor that makes its way back to the compressor inlet (suction side) to be compressed back into a liquid and start the journey all over again.

This is the basic principal of nearly all air conditioning systems. The cold evaporator coil is used to make building air cold and the hot condenser coil is used to release the absorbed heat to the outside. Air conditioners don't really create any cold, they just absorb heat from the inside air and expel that heat to the outside air.

HEAT PUMPS

Heat pumps are air-conditioners run in reverse. Many heat pumps can be used to both cool and heat building air. Heat pumps work by using a traditional refrigerant circuit with one added component, a reversing valve. A reversing valve is an electronically operated valve that effectively switches the positions of the evaporator coil and condenser coil in the refrigeration loop.

With the reversing valve in the normal, or cooling position, the refrigeration loop works exactly as described above and the system will cool the building as expected. With the reversing valve in the reversed or heating position, the evaporator coil inside the building becomes the hot condenser coil and the condenser coil outside the building becomes the cold evaporator coil. In the reversed mode, a heat pump will absorb heat from the outside air and release that heat into the building to warm the building space.

As long as the outside air is warm enough for the heat pump to

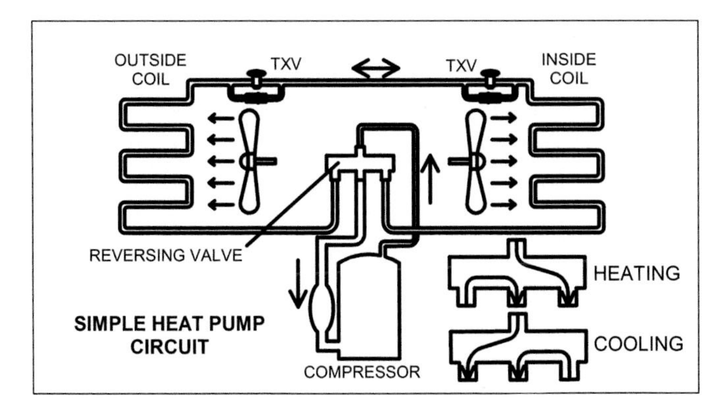

Figure 7-3. Components of a simple heat pump. Notice how the reversing valve effectively swaps the two coils.

be able to absorb heat, usually around 40°F, heat pumps are a more efficient source of heat than either electric heat strips or combustion heating equipment. However, once outside temperatures drop below this temperature, another source of heat must be available. Often electric heat strips are included as supplemental heat. Because of this limitation, heat pumps may not be an economical choice in cold climates.

Heat pumps are more efficient than other types of heat because heat pumps do not need to produce heat, they only move heat from one place to another. Other types of heating equipment produce heat by burning gas or oil or by using electric current to heat a high resistance wire. This approach converts fuel or electricity to heat. Heat pumps don't do this. Heat pumps move heat from one place to another. It takes much less energy to move heat that is already there.

Coefficient of performance (COP) is used to compare the efficiency of different types of heating equipment. Electric resistance heaters have a COP of 1.0 since one unit of electricity is converted to one unit of heat. A COP of 1 means the efficiency is 100%. Heat in equals heat out. Heat pumps can have a COP between 2.8 and 4 since they can have a heat output that is 2.8 to 4 times as large as the electricity they use. This makes heat pumps a very cost effective choice in most climates.

PREVENTIVE MAINTENANCE OF AIR CONDITIONING AND REFRIGERATION EQUIPMENT

A common air conditioner configuration is for the condenser coil, compressor, and condenser fan to be installed outside the building as one unit and for the evaporator coil, expansion valve and evaporator blower to be installed inside the building. This type of system is called a split system. The outside equipment is simply called the condenser unit and the inside unit is usually called an air handler or DX unit for direct exchange.

Another common configuration is for all of the parts of the air conditioning circuit are installed in one unit, this is known as a package system. A package system can be installed inside the building or outside. When a package unit is installed outside, building air is taken to the unit to be conditioned through ductwork. When a package unit is installed inside a building, hot air is discharged to the outside through ductwork.

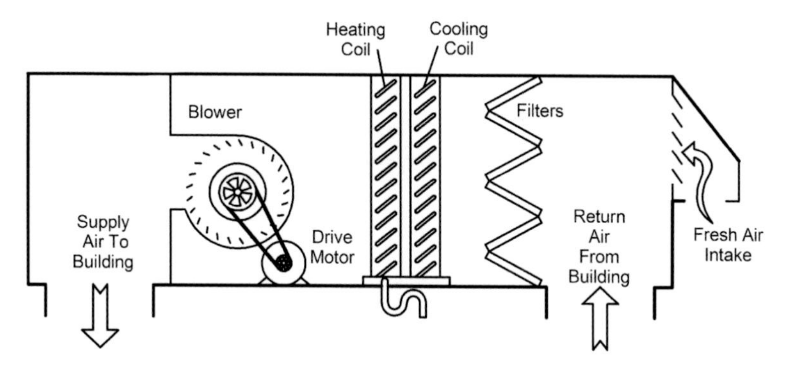

Figure 7-4. Components of a typical air handler unit.

In either case, the condenser equipment is kept separate from the air handling equipment, even when they're installed in the same cabinet. The condenser side of the system is responsible for removing heat from the system to the outside air and the air handler side of the system is responsible for cooling the air and delivering the cooled air to occupied spaces.

The air handling side of the system can also include heating equipment, filters to clean the air, outside air dampers or economizers to allow fresh air to enter the building, humidifiers, and other equipment.

Preventive Maintenance of Condensing Units—Compressors

With the exception of large chillers providing hundreds of tons of cooling, most air-conditioning and refrigeration equipment uses reciprocating type compressors. These have pistons that move up and down and compress refrigerant with each piston stroke.

The compressors of small air-conditioning and refrigeration equipment are often hermetically sealed inside a welded canister and are not serviceable. If a hermetically sealed compressor fails, it is simply cut out of the refrigeration circuit (after the refrigerant is recovered from the system) and a new compressor is brazed in its place. Larger air conditioning equipment, usually over 7 or 10 tons cooling capacity, may use compressors that can be serviced. These semi-hermetic compressors can be disassembled and repaired on-site rather than discarded and replaced. The current trend is for larger cooling units to use multiple smaller hermetic compressors and as a result semi-hermetic compressors are becoming less common.

Figure 7-5. Hermetic compressors are completely sealed and have no serviceable parts.

Like most rotating machinery, the most important part of refrigeration compressor preventive maintenance is lubrication. Many serviceable refrigeration compressors have oil sight glasses where the oil level in the compressor can be checked visually. The oil line should be visible on the sight glass. If not, oil can be added to the system using a pressurized oil pump. This needs to be done by someone with the proper EPA certification to work on the closed refrigerant system. Unless the closed refrigerant loop has experienced a refrigerant leak, the oil level in the system should not change.

Refrigerant oil selection is critical. Not all refrigeration oils are compatible with all types of refrigerant and not all refrigeration oils can be mixed with the oils already in the system. There are an enormous number of mineral and synthetic refrigerant oils and the wrong oil will not blend with the refrigerant and will cause oil starvation and compressor failure. Always refer to the manufacturer's requirements on oil selection.

Figure 7-6. Semi-hermetic compressors can be disassembled for repair. Notice the oil sight glass on the bottom right for checking the oil level.

Preventive Maintenance of Condensing Units—The Condenser Coil

Condenser Coils need to be kept clean. Dirt blocks air flow and reduces the rate that heat can flow out of the coil into the outside environment. Dirty coils make refrigeration equipment inefficient and can cause a system to overheat and shut down all together. If the aluminum cooling fins are bent over or crushed, blocking air flow, a fin comb should be used to straighten the fins.

Condenser coils are usually not protected by any type of filter. Condenser coils need to be cleaned annually, at the start of the cooling season. There are a variety of cleaners available that are sprayed on the coils and hosed off with water several minutes later. Condenser coils located outside on the ground (hopefully on a concrete or plastic pad) should have all leaves and other debris removed from inside the condenser unit and all grass, weeds, and shrubbery needs to be kept and at least 18″ away from coils to allow for enough air flow around the coil.

Preventive Maintenance of Air Handlers—The Evaporator Coil

Evaporator coils (the inside coil) is protected by a filter. If a filter is blocked with dirt and dust, air will start to find a path around the coil. When this happens, the dirt in the air bypassing the filter will clog the evaporator coil. It is important that air conditioner filters are changed frequently. Operating environment, hours of operation, and other factors will effect the necessary frequency of filter changes but most organizations change filters quarterly.

If filters are kept clean, evaporator coils need little maintenance. Air handler cabinets should be opened annually, at the start of the cooling season. Check the coil for dirt build up and use a fin comb should to remove any dirt or lint that has built up on the intake side of the coil.

Liquid coil cleaners are available but probably aren't needed and can create a housekeeping mess since many of the cleaners foam up and drip off of the coil missing the condensate pan.

Figure 7-7. A quick visual inspection of the entire unit when performing PM will catch obvious problems like a broken condensate drain trap.

Preventive Maintenance of Air Handlers—Air Filters

The primary function of an air filter is to protect heating and cooling coils from becoming clogged with dirt and dust present in the air stream. For many years, protecting coils from clogging was the only consideration when selecting and installing air filters. Since small particles and fibers will pass through coils without clogging them, "furnace filters" were designed to only catch the largest particles. Since the air passing through the filter and into the building was building air mixed with some fresh outside air, it was at least as clean as the outside air. At the time, this was considered sufficient filtering.

Today's workforce is concerned about indoor air quality (IAQ), sick building syndrome, allergens, and other health concerns related to air contaminants. Indoor air that is as clean as the air outside is no longer considered acceptable and air filters need to eliminate more and smaller dust particles. The facilities manager or maintenance supervisor needs to understand how filter efficiency is measured and how it relates to removing contaminants from the air.

There are presently three systems that manufacturers use to determine and specify how well a filter removes dust particles from the air. These three systems were established by the American Society of Heating, Refrigeration, and Air Conditioning Engineers as ASHRAE standards 52.1 and 52.2.

The first rating system is the dust weight arrestance rating which indicates how well a filter captures dust particles and fibers larger than 10 microns. It is used to evaluate fiberglass mat type furnace filters that are intended to protect heating and cooling coils but are not very effective on fine dusts, mold spores, pollen, or other tiny allergens which may contribute to IAQ problems.

The second filter efficiency rating system is the dust spot efficiency rating which indicates how well a filter captures smaller particulates (.3-6 microns) such as fine dusts and smokes. This test is usually used to evaluate pleated and fabric bag type filters. These filters do remove much of the fine particles which contribute to IAQ difficulties.

The third and newest filter efficiency rating system is the MERV or minimum efficiency reporting value. This test measures a filters effectiveness at collecting both small and large particles (.3-10 microns) and has been the established standard since 1999. However, all three standards are still in use.

Since all three standards measure the efficiency of a filter to collect particles of different sizes, there is no way to directly compare one rating system to another. Figure 7-8 shows some guidelines for comparison but since the three different systems measure different performance characteristics; there is no way to convert one system to another. Use Figure 7-8 only as a rough guide for comparison.

When choosing filters, more filtration is not necessarily better. Higher filter efficiency means reduced air flow, greater pressure drop across the filter, and more frequent filter changes since higher efficiency filters will load up with dust faster. Higher efficiency filers also cost more. Filters with MERV ratings over 8 may require modifications to ductwork to allow larger filter banks for sufficient air flow.

Most commercial office buildings, schools, and similar facilities should be using air filters with MERV ratings between five and eight. Under most conditions, filters with these ratings will need to be changed roughly four times each year.

Preventive Maintenance of Air Handlers—Condensate Equipment

Humidity in the air will condense on the cold surface of evaporator coils. For larger air conditioning equipment, this condensate can amount to several gallons an hour in humid conditions. This is good since one of the functions of air-conditioning is humidity removal, but something needs to be done with all that water.

In most cases, there is a galvanized steel or plastic tray under the evaporator coil that catches the condensate dripping off the coil.

Typical Air Filter Type ▼	Disposable Panel Filters, Fiberglass & Synthetic Filters, Permanent Self Cleaning Filters, Electrostatic Filters, Washable Metal Foam Filters	Pleated Filters, Extended Surface Filters, Media Panel Filters	Non-Supported Bag Filters, Rigid Box Filters, Rigid Cell / Cartridge Filters	Non-Supported Bag Filters, Rigid Box Filters, Rigid Cell / Cartridge Filters	HEPA Filters, ULPA Filters, SULPA Filters
MERV (Std. 52.2)	1-4	5-8	9-12	13-16	17-20
Average Dust Spot Efficiency	<20 %	<20 to 35%	40 to 75%	80 to 95%+	99.97% 99.99% 99.999%
Average Arrestance ASHRAE Std. 52.1	60 to 80%	80 to 95%	>95 to 98%	98 to 99%	N/A
Particle Size Ranges	>10.0 microns	3.0-10.0 microns	1.0-3.0 microns	0.30-1.0 microns	<0.30 microns
Typical Air Filter Applications	Residential, Light Commercial, Equipment Protection	Industrial Workplace, Commercial, Paint Booths	Industrial Workplace, High End Commercial Buildings	Smoke Removal, General Surgery, Hospitals and Health Care	Clean Rooms, High Risk Surgery, Hazardous Materials

Figure 7-8. Comparison of various filter rating systems

Condensate trays or pans need to be inspected annually, at the start of the cooling season, to be sure the tray is sloped toward the drain, to make sure the tray is not growing algae, to check steel condensate trays for rust or holes, and to check the tray for foreign matter and dirt which could clog the condensate drain line.

Dirty condensate trays can be rinsed clean with a garden hose. Algae control tablets can be added to the tray to prevent algae growth, and epoxy sealers can be applied to repair any rust or corrosion holes.

Because condensate flow is so slow in the condensate drain lines, condensate lines tend to clog easily. The slow dripping of condensate simply isn't fast enough to flush all of the debris from the drain line. Drain lines should be blown out with compressed air at the start of the cooling season. This can be done by blowing air from the air handler end of the drain line or from the end that is opened to the outside or to the building's sanitary drains.

When an air handler (and the incorporated evaporator coil) is located in a basement or otherwise below the level of condensate discharge, a condensate pump is used to pump condensate water up to the proper location. A failed condensate pump will cause a minor flood. Condensate pumps should be tested quarterly. Either pour water into the condensate pump or operate the float switch manually to be sure the pump operates properly. Watch the water level in the sump to be sure it goes down when the pump operates.

Preventive Maintenance of Air Handlers—Blowers

Every air handler uses a fan to move conditioned air to occupied spaces. These fans are usually centrifugal fans, also called "squirrel cage fans." Fan blades need to be kept clean, dirt on fan blades can greatly effect the efficiency of the fan and therefore how much air the fan can move. Fans should be inspected twice annually, at the start of the cooling and heating seasons, and cleaned when needed.

Bearings need to be oiled or greased quarterly or according to the manufacturer's recommendations. Belt drives should be checked quarterly and tension and alignment adjustments made and worn belts replaced. It's a good idea to stock at least one set of belts for each air handler. This will save time having to return to replace belts at a later date if a problem is discovered.

**Preventive Maintenance of Air Handlers—
Heating and Other Equipment**

Air handlers can include many types of heating equipment. Usually the heating equipment is in series with the cooling coil and air is drawn by the blower motor through both the heating and cooling equipment during all seasons. A lockout circuit prevents both heating and cooling equipment from working at the same time unless wintertime humidity control is a problem. In that case, heating and cooling equipment may operate together.

The most common problem with electric heat is burned out heating elements or blown fuses serving the individual elements. At the start of each heating season, the heat strips should be visually inspected and your HVAC service tech or electrician should take an amperage reading of each heat strip in operation. This will reveal any heat strips that are not functioning or heat strips that are drawing more than their rated amperage and likely to fail.

Gas heating equipment should be serviced annually, at the start of the heating system. Burners should be cleaned, gas pressure checked, igniters or pilot flames tested and the operation of flame sensors verified. The burner flame should be adjusted for highest efficiency. Finally the heat exchanger needs to be inspected for any cracks to make sure that combustion gasses, including carbon monoxide, aren't getting into the building air stream.

Oil burners should have nozzles replaced annually. Igniters and flame sensors need to be tested for operation. Oil filters should be changed annually and the oil pump pressure verified. Combustion gasses should be analyzed to determine burner efficiency. Finally the heat exchanger needs to be inspected for any cracks to make sure that combustion gasses, including carbon monoxide, aren't getting into the building.

Many air handlers will include a heating coil in addition to the evaporator cooling coil. This heating coil can be a hot water or steam coil which is provided with hot water or steam from a separate boiler plant located elsewhere in the building. The heating and cooling coils can be stacked on top of each other or even designed as one coil assembly with two sets of pipes, one for cooling and one for heating. Heating coils require the same inspection and cleaning schedule as evaporator coils discussed earlier.

COOLING TOWERS AND COOLING LOOPS

So far, we've only discussed air-conditioning equipment that exchanges heat directly with the air. Usually by absorbing heat from the air inside the building and expelling that heat into the air outside the building. Larger HVAC systems often use water as a heat exchange medium and use pipes and pumps to move the heated or chilled water from one place to another. Water can absorb and hold more heat than air and it is more efficient to exchange heat with water.

There are so many variations of using water as a heating or cooling medium that including all the different types would be impossible but we'll discuss some of the more common variations.

The most common method of using water to exchange heat with the outside air is a cooling loop leading to an evaporative cooling tower. A refrigerant condenser coil is often placed in a bath of water and the water is used to absorb heat from the condenser coil to cool the refrigerant. The hot (or at least warm) water is then pumped to a cooling tower outside of the building where it flows under gravity down a series of screens while cooling tower fans draw air through the hot water. Some of the hot water evaporates taking heat away from the remaining water leaving cool water. The cool water is pumped back into the building to cool the condenser again. The water loop that cools the condenser coil is called the condensate water loop which should not be confused with condensate water that drips from an air source evaporator coil. There are four common types of cooling towers and a cooling tower that has the condensate water open to the atmosphere as discussed above is often called a *wet cooling tower.*

Cooling towers can also have a closed water coil similar to a refrigerant condenser or evaporator coil. Fans draw air through the coil to remove heat from the water inside. With this type of cooling tower, no water is evaporated or open to the outside air. This type of cooling tower is often called a *dry cooling tower.*

A third type of cooling tower is a hybrid of the wet and dry cooling towers. This *hybrid cooling tower* has a closed coil as with the dry type tower but also has its own source of open water that is sprayed over the closed coil to provide more efficient cooling than air alone.

There is a fourth type of cooling tower that does not use a water loop at all called an *evaporative condenser.* An evaporative condenser is similar to the hybrid cooling tower example above except that refriger-

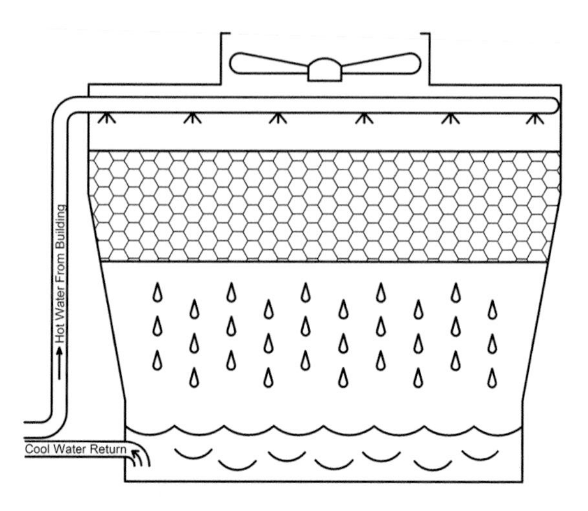

Figure 7-9. Water in a wet cooling tower cascades down through a series of screens while air is drawn across the water to remove heat. The cooled water is returned to the building.

Figure 7-10. This hybrid cooling tower has a closed condensate loop that flows through a closed cooling coil. A separate source of water flows over and cools the water in the coil.

ant flows through the coil within the cooling tower. Instead of using a water loop to transfer heat from the building to the cooling tower, the refrigerant condenser coil is located within the cooling tower. Water is sprayed over the surface of the refrigerant condenser coil. The evaporation of water in the cooling tower absorbs heat directly from the refrigerant more efficiently than a traditional air cooled condenser coil.

In most HVAC systems using cooling towers or central chiller plants, the cooled water is circulated throughout the entire building and circulated through cooling coils or used to cool condenser coils of individual air conditioning units serving each area of the building. Boilers can be part of the same water loop to provide warm water as a source of heat for individual heat pumps or water source heating coils for each location.

As an example of how complicated these systems can become I have worked in a building that used 4 separate water loops to transfer heat and one refrigerant loop. A drawing of this system is shown in Figure 7-11. This particular system used more than 40 individual water-source heat pumps throughout the building. Each of these heat pumps had its own refrigerant loop and exchanged heat through a condenser coil immersed in a water loop that ran through the building

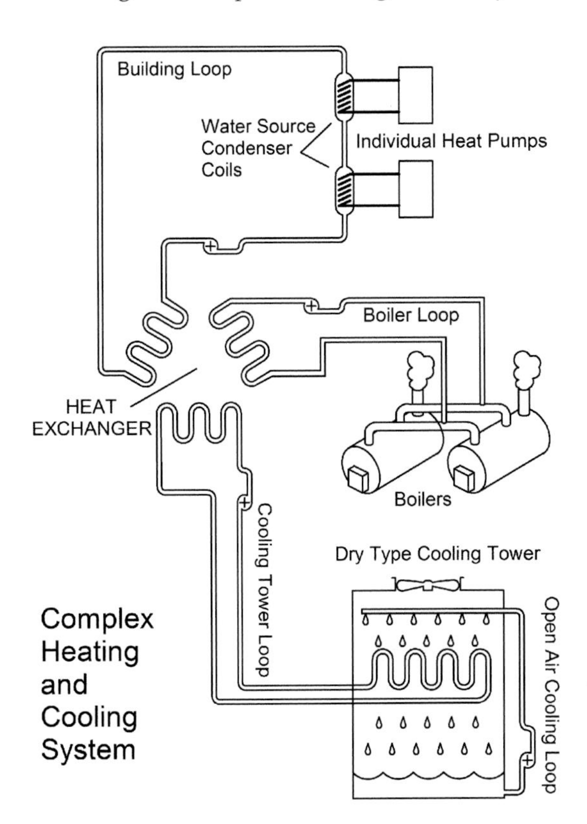

Figure 7-11. A complex heating and cooling system utilizing four separate water loops to transfer heat from one component of the system to another.

to each heat pump. This building loop exchanged heat with a cooling tower water loop in the summer or with a boiler heated water loop in the winter through a plate type heat exchanger. The cooling tower was of the dry type and had its own internal water loop to remove heat from the cooling tower loop between the cooling tower and heat exchanger.

Preventive Maintenance of Cooling Towers and Cooling Loops

PM of cooling towers is straightforward once you understand the separate parts that make a cooling tower work. Spray pumps should be greased quarterly when the cooling tower is in operation. Cooling tower fans and motors should be greased quarterly and belt drives checked and adjusted quarterly. Spray nozzles can clog and should be inspected monthly and cleaned or replaced as needed during cooling tower operation. During the monthly inspection, verify operation of any automatic makeup water valves. Cooling towers should either be drained for the winter to prevent ice damage or should have water heaters that maintain the water above freezing temperature. These heating elements should be checked prior to the cold season. Leaves and other debris should be removed from the cooling tower sump monthly or more often if clogging is an issue. Once each year, the tower should be drained and the water pan checked for rust or corrosion. Epoxy coatings are available to repair minor damage and leaks to cooling tower pans and there are companies that specialize in doing this work.

The only maintenance that is required for cooling loops is chemical treatment. Chemicals to prevent corrosion and scale buildup should be added to the closed cooling loops, either by automatic treatment equipment or manually by your service company. All systems are different and there are many chemical treatment options available. A local chemical treatment company or your HVAC service company can best help you decide what chemical treatment options are best for your facility.

CHILLERS

Very large facilities often rely on chillers instead of conventional air conditioners for their cooling. Chillers are often sized to provide hundreds of tons of cooling capacity in one relatively compact machine. One or two large chiller plants is often used to provide cooling for an entire building or even an entire campus of buildings.

Chillers are large self contained water chilling machines with all of the parts of the refrigeration machinery built into one unit. Heat from the building can be exchanged directly with the outside air or can be exchanged through a water loop and cooling tower. In either design, chillers cool a large volume of water which can be circulated throughout the building to provide space cooling via water source duct coils, or unit ventilators.

Chillers operate like conventional air conditioners with one notable exception. The refrigerant compressor is often of a unique design. Like conventional air conditioners many smaller chiller plants use reciprocating refrigerant compressors, often several compressors on one machine. However, larger capacity machines often compress the refrigerant using a high-speed centrifugal compressor with an impeller which can rotate as fast as 50,000 RPMs compressing the refrigerant through centrifugal force.

Sophisticated onboard computers manage every detail of chiller operation. Compressor speed and staging, condenser and cooling water flow, water temperatures, refrigerant pressures and temperatures, expansion valve position, and many other items are managed constantly to

Figure 7-12. Centrifugal chillers are complex and expensive machines and should be PMd by factory-trained personnel.

ensure energy efficient operation.

Every manufacturer has different maintenance procedures for their chillers and there is probably no other single piece of equipment in your building, with the possible exception of the roof, where improper maintenance could be more costly. In one example from my work, A 250 ton chiller most likely had an incorrect expansion valve adjustment. This allowed a small amount of liquid refrigerant to enter the compressor while the impeller was spinning at full speed. When the rotating impeller struck the liquid refrigerant at high speed, the impeller shattered leaving shards of metal which were pumped throughout the machine along with the refrigerant and oil. The repair cost was over $40,000.

Manufacturer's maintenance standards must be followed, preferably by factory trained and certified personnel. I am a big believer in a skilled maintenance department handling nearly every project in-house. Chillers are the exception. Skilled in-house HVAC personnel can handle basic repairs to equipment related to the chiller such as circulating pumps, flow switches, chemical treatment systems, and other items. The chiller itself should be maintained by someone who is an expert in your particular brand of chiller.

AUTOMATED CONTROLS

HVAC systems are almost certainly the most complex systems found in any commercial building. In the recent past, the temperature of each part of a building was controlled by individual stand alone air-conditioning units. In many buildings, hot water or steam was piped throughout the building and there was no individual temperature regulation for each area. Either the heat was on for the building or it was off. Cooling was often controlled locally because ac units were small but there were no provisions for humidity control or any strategy to make the HVAC systems energy efficient.

By contrast, in most of today's buildings, all of the heaters, air conditioners, exhaust fans, variable frequency drives, circulating pumps, outside air dampers, cooling towers, and a network of temperature, humidity, and occupancy sensors work together, as one large system to monitor and control every aspect of a building's environment. Having such complex HVAC systems allows us to provide more comfortable interior conditions then ever before while wasting less energy.

The downside of HVAC systems that utilize these direct digital controls (DDC) or building automation controls (BAC) is that they limit the amount of repair work that can be done in-house. Most HVAC systems require outside contractors to maintain the digital control portion of the system. This small problem doesn't mean that automated HVAC controls are a bad thing. A good building automation system will save much more in energy costs than will be spent on competent service people to maintain the system's original operating parameters.

The most common problem with networked or integrated HVAC systems is that they are constantly being "tweaked" by in-house staff to the point that they no longer function as designed. Small adjustments and minor changes to the system are made by well meaning maintenance workers. Over several years, so many small changes are made that the system can no longer do its job as it was designed.

The maintenance department might be asked to adjust an air damper to satisfy an office worker who doesn't like to feel a draft. A well intentioned maintenance mechanic might disable an outside air damper to prevent rain from blowing into the building. We might rewire a networked AC unit to operate from a wall thermostat if a company executive complains about not having direct control of the temperature in her office. Add up these small modifications that are done over the years, and it's easy to see why so many building automation systems end up as expensive and complicated disasters.

Automated environmental control systems are good for buildings. They have been around for a couple of decades and are getting smarter all the time. They do an excellent job of keeping occupants happy while conserving as much energy as possible. However, they are fragile systems and must be maintained properly by people who understand the intended operation and design of the entire system.

One of the most effective ways to maintain the original intended operation of automated controls is to write a controls narrative which briefly and simply describes the operation of the entire system. This narrative can be posted permanently in boiler rooms, mechanical rooms, and inside equipment cabinet doors. The narrative should explain the function of all the system components and explain what environmental conditions trigger what type of system response.

With a controls narrative available to everyone who works on the system, it is more likely the original intended system operation will be maintained.

HVAC PM CHECKLIST

The following checklist of items will help you when setting up your preventive maintenance schedule.

- ❑ Clean condenser coils annually at the start of the heating season
- ❑ Keep all vegetation at least 18″ from condenser coil
- ❑ Change air handler filters quarterly
- ❑ Check evaporator coils annually at start of cooling season for dirt.
- ❑ Check condensate trays, treat tray with algae control tablets, and clean condensate drain lines annually
- ❑ Test condensate pumps monthly
- ❑ Inspect air handler blower fans, belts, and motors twice annually.
- ❑ Oil or grease blower motors quarterly
- ❑ Take ohm readings of electric heat strips annually at start of heating season. Replace any defective elements or blown fuses
- ❑ Have gas or oil fired heating equipment serviced professionally annually at the start of the heating season
- ❑ All circulating pump motors need to be greased or oiled quarterly
- ❑ Inspect cooling tower drive belts quarterly
- ❑ Grease or oil cooling tower fan motors quarterly
- ❑ Inspect, clean, or replace spray nozzles on cooling towers monthly during cooling season
- ❑ Clean leaves or debris from the cooling tower sump monthly and check the operation of the sump makeup water valve
- ❑ Cooling tower heating elements should be tested annually at the start of winter
- ❑ Cooling towers should be drained annually to inspect the sump tank for corrosion or rust

CHAPTER 7 SUMMARY

- The HVAC system is one of the most expensive and complex systems found in any building.
- Anyone working on the refrigerant circuit must be certified by to EPA to work with refrigerants.

- The complexity of HVAC systems often makes it necessary to rely on outside contractors to maintain and service HVAC equipment.
- The design of air conditioning equipment has not changed significantly since its invention by Willis Haviland Carrier in 1902.
- Cooling by a refrigeration circuit is accomplished by evaporating a refrigerant with an evaporator coil. The evaporating refrigerant absorbs heat from the surrounding air cooling the air.
- The absorbed heat is then transferred with the refrigerant to a condenser coil where the refrigerant is condensed back into a liquid thereby giving up its heat to the outside air.
- Heat pumps are air conditioners run in reverse. Instead of absorbing heat from the interior space and releasing it to the outside environment, heat pumps absorb heat from outside and release the heat inside.
- Selecting the proper refrigerant oil is critical. The wrong type of oil may not mix with a particular refrigerant resulting in compressor failure.
- Many HVAC systems rely on water to transfer heat from one location to another. Oftentimes, multiple water loops are used to make several transfers.
- Large facilities may use chiller plants to produce chilled water which is used to cool the building space. Chillers are very expensive and complex pieces of equipment and are best maintained by factory authorized service personnel.
- Today's buildings use automated building controls to monitor and regulate all aspects of HVAC within a facility. These systems allow a level of building control and energy efficiency that would not otherwise be possible. The down side to these systems is they must be maintained by technicians with specialized knowledge and skills.

Chapter 8

Belt Drives

The preventive maintenance of buildings involves many types of equipment that rely on flexible drive belts for their operation. Air handlers, exhaust fans, and rooftop air-conditioning units (RTUs), air compressors, and other equipment utilize drive belts in most facilities.

Drive belts are such a simple device that including a technical section on them may seem unnecessary. The fact is unsatisfactory installation, improper belt selection, and poor adjustments make belt drives account for a very high percentage of preventable mechanical failures in facilities. By having a better understanding of some of the engineering of drive belts and by taking the time to install and adjust them correctly, many mechanical failures can be avoided. And avoiding mechanical failures is why we're performing PM in the first place.

Drive belts are commonly called "vee-belts" or sometimes "Gates belts" after John Gates who developed the V drive belt in 1917. V-shaped

Figure 8-1. Belt-driven blower

drive belts solved two problems that were associated with flat belts in use at the time. The first problem was that flat belts could slide sideways on the flat pulleys or sheaves causing alignment problems. This was solved because vee-belts can't slide out of the v-shaped groove on their pulleys. The second problem was that flat belts often slip under heavy load. vee-belts tend to wedge themselves into the pulley groove under load increasing the resistance to slipping.

THE 3 TYPES OF DRIVE BELTS

Most of the drive belts that are used in equipment in our buildings are going to be one of three types. The three different types are "classic cross section" belts, "narrow profile" belts, and "fractional horsepower" belts. Figure 8-2 shows the classic profiles in the top row, the narrow profile belts in the middle row, and the fractional horsepower belts in the bottom row.

The first two belt types, the "classic cross section" belts and the "narrow profile" belts are known as industrial belts and are typically

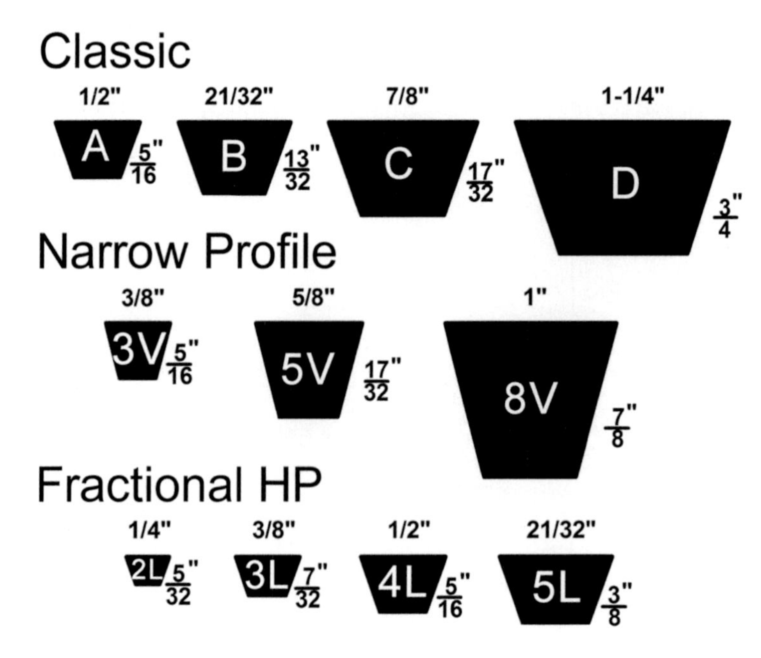

Figure 8-2. The three-drive belt types

used on equipment with power requirements over 1 horsepower. The profiles of drive belts are standardized. and the classic cross section belts are designated by the letters A, B, C, and D. Narrow profile belts are narrower than classic belts and are designated using the letters 3V, 5V, and 8V.

The third type of belt is the fractional horsepower belt. These are usually used on equipment with power requirements below one horsepower, although there is quite a bit of cross over between fractional horsepower belts and industrial belts in equipment around 1hp. Fractional horsepower belts are designated by the letters 2L, 3L, 4L, and 5L.

DRIVE BELT LENGTHS

Now that we understand the cross section designations of drive belts, we need a way to determine the lengths of drive belts.

There is a standard numbering system for drive belts that identifies both the cross section shape and the length of the belt. Drive belt lengths are measured in inches. Both the Fractional horsepower belts (cross sections 2L, 3L, 4L, and 5L) and V-series belts (3V, 5V, and 8V) are numbered with the cross section designation followed by the length in inches plus a zero. For example, a fractional horse power belt with a 2L cross section that was 22 inches in length would be a 2L220 belt. A 3V550 belt would have a 3V profile and be 55 inches in length.

Industrial belts of cross sections A, B, C, and D have a slightly more complicated numbering system. Like the fractional Hp belts and the V-series belts, the first part of the belt number is the cross section letter. The second part of the belt number will also give you the belt length in inches but you will need to do some simple addition. For the A cross section belts, you need to add two inches to the part number to get the length of the belt. For example, a drive belt with the part number A43 is 45 inches in length since 43 + 2 = 45. In the same way, you need to add 3 inches to a B cross section belts and 4 inches to the C cross section belt. A B35 belt would therefore be 38″ in length (35+3) and a C96 belt would be 100 inches all the way around (96+4).

Measuring the length of a drive belt is not as simple as it seems. Since the top and bottom surfaces of a drive belt stretch and compress as the belt bends around the pulleys, the measurement of length is not actually taken at the outside (top) or inside (bottom) surface of the belt. The actual length measurement of a drive belt is taken at a location known as the "neutral axis" which is usually where the reinforcing cords are located within the belt. Figure 8-3 shows the reinforcing cords at the neutral axis inside a drivebelt.

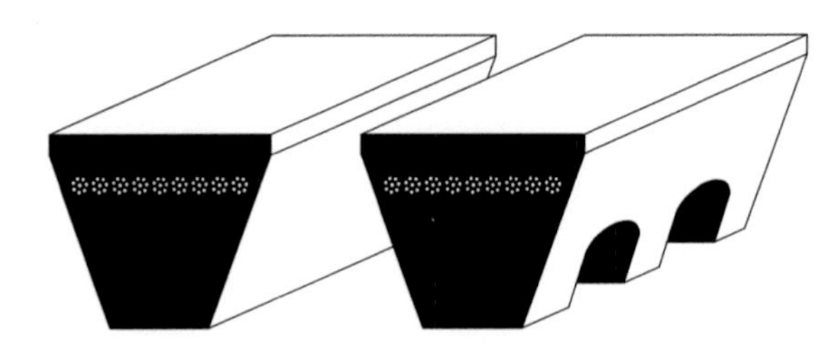

Figure 8-3. Reinforcing cords are located at the neutral axis.

It can be almost impossible to accurately measure the length of a drive belt. Drive belts are just too flimsy to get a good measurement. Most industrial suppliers will have a tool for measuring belt length. The tool is a pair of sliding pulley's mounted on a rule. A belt is wrapped around the pulleys and the pulleys are slid apart to tighten the belt. The belt length is then read directly from the measurements marked on the rule. If you can't find a belt's size printed on an old belt and you don't have the proper tool for measuring the belt, the easiest way to measure the length is to cut the belt and stretch it out flat.

WHAT IF I DON'T HAVE AN OLD BELT TO MEASURE?

Sometimes we don't have an old belt to measure because the old belt is too damaged or even missing. If we don't have an equipment owner's manual or the old belt, it's still not too hard to figure out what size belt will fit a particular piece of equipment. First, we can figure out the cross section of the belt by trying some spare belts and second, we can determine the length of the required belt with a little math.

The first step to figuring out the size of a missing belt is to determine the cross section of the belt that fits the pulleys. This is best done by trying a few spare belts of different cross sections in the pulley groove to see what fits. Remember that most of the time, the top of the belt should just about sit flush with the top of the pulley groove and that

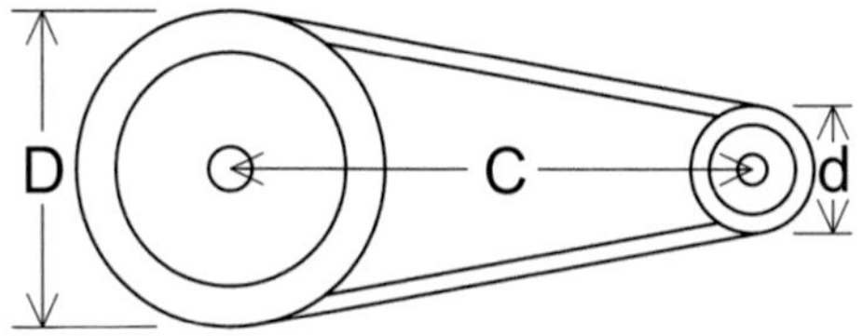

$$L = 2C + 1.57(D+d) + \frac{(D-d)^2}{4C}$$

Figure 8-4. Formula for determining drive belt length

the belt should never bottom out and touch the bottom of the groove.

When trying to figure out what cross section fits best, it's also useful to remember that the classic cross section belts (A,B,C,D) and the fractional horsepower belts (2L,3L,4L,5L) belts are much more common than the less often seen narrow profile belts (3V,5V,8V). Most of the time it will be easy to see what profile fits the best. The two belt profiles that can be difficult to tell apart are the A profile classic section belt and the 4L profile fractional Hp since these are almost exactly the same. If these two belts are difficult to tell apart, don't worry. These are the only two belt profiles that are interchangeable so you can choose either one and still be right.

Once we've figured out the cross section of the belt we need, the second part of figuring out the size of a missing belt is to do some math to figure out the length of the belt. To do this, we need to know 3 things: First, the diameter of the larger pulley; second, the diameter of the smaller pulley; and third, the center to center distance of the pulley shafts. Figure 8-4 shows two pulleys and the formula used to calculate the length of belt needed.

OTHER DRIVE BELT CHARACTERISTICS

Standard drive belts are available with a few options. One of these optional styles is notched belts. A notched belt has notches molded into the bottom surface of the belt and is sometimes referred to as a cogged belt. Notched belts can be flexed tighter to run on smaller diameter pulleys than standard drive belts. Notched belts will have an "X" added to the part number after the cross section numbers. 5VX, 3VX, and AX are notched belts in 5V, 3V, and A cross sections.

Some of the larger horse power motors will use several drive belts side by side on pulleys with multiple grooves. When this is the case, the belts need to be ordered as "matched." That means the belts have been verified to be the same length so that all the drive belts will have the same tension when installed. Ordering matched belts is not as much a problem today as it was 20 years ago. Manufacturing tolerances in belts have improved and most belts are close enough to standards to be considered matched right off-the-shelf. It's still a good idea to tell your supplier if you need matched belts to avoid any problems.

Lots of drive belt types exist other than the ones discussed above. Some examples include metric belts, synchronous drive belts, flat belts, grooved belts, and double V belts that can run on both the inside and outside surfaces of the belt. It is rare for any of these types to be used in the types of equipment that are encountered during PM of most facilities. Even when an equipment manufacturer has their own part number for a proprietary drive belt, it probably matches exactly with one of the standard industrial or fractional belts available, and often at a fraction of the price of the manufacturer's original equipment.

PROPER OPERATION

Drive belts in continuous operation should be expected to have a useful life of about one year if installed and adjusted correctly. Proper installation means well aligned pulleys and proper belt tension.

Figure 8-5. A pair of matched belts for a high-torque application

Pulley alignment can be checked by using a straight edge such as a steel ruler or steel carpenter's square or a piece of string held taught across the face of the pulleys. If pulleys are found to be out of alignment, the time spent making adjustments to the motor orientation or to the pulley's depth on the motor shaft will save time replacing damaged or thrown belts later.

Drive belt tension can be checked using a quick rule of thumb. A properly tensioned belt should deflect approximately 1/64 of its span length when deflected by hand. For multiple belts, each belt should be deflected individually instead of attempting to check all of the belts at once. Belts that are too loose will slip and wear prematurely. Slipping belts can usually be heard squealing at startup. Belts that are too tight will also wear prematurely from excessive friction as the belts enter and leave the pulley groove during pulley rotation.

Although rarely done, when installing new drive belts, it is good practice to run the belts for a day to give the belts an opportunity to stretch and then re-tension the belts again. Belt tension should be checked, and adjusted if needed, each time a machine is maintained.

PROBLEMS WITH BELT DRIVES

Pulleys that are misaligned or shafts that are misaligned will cause premature wear and overheating to belts. Misaligned pulleys cause side pressure and twisting of belts which will cause wear to the belt edges, hardening caused by overheating, and fatigue cracks due to excessive flexing. One of the tell-tale signs of belts with worn edges is a belt that sits too low in the pulley groove. The bottom surface of a belt should never touch the base of the pulley groove and the top surface of a drive belt should barely stick up above the rim of the pulley.

Since the metal pulleys wear so much slower than rubber belts, most repairs involve replacing drive belts and not the pulleys. However, after years of operation pulleys can become worn on the inner surface of the belt groove. The easiest way to find this is to look at the groove walls on the inside of the pulley groove. A worn pulley will have concave walls while a new pulley will have perfectly flat sides inside the belt groove. When the side of the pulley groove is worn down, the drive belts will sit too deeply in the pulley groove.

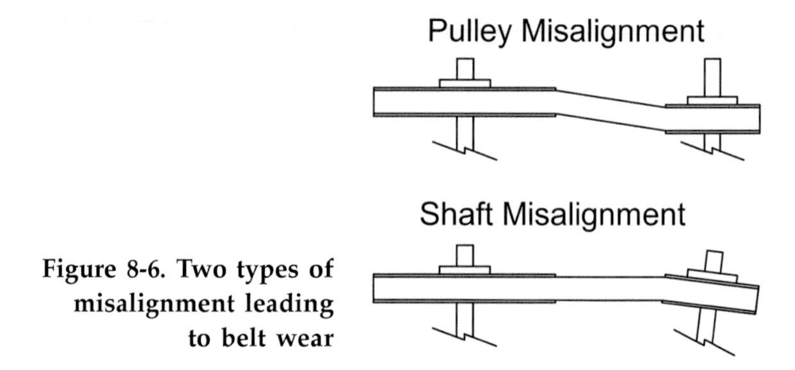

Figure 8-6. Two types of misalignment leading to belt wear

Once a pulley (or belt) is worn to the point that the belt is sitting on the bottom surface of the belt, the belt can no longer wedge itself into the groove and excessive belt slipping will occur. The desirable friction between a pulley and drive belt is created when the belt is squeezed between the sides of the pulley groove. Once a drive belt has bottomed out in the pulley groove, it is no longer being squeezed and the needed friction is gone. Many times, worn drive belts appear to be in perfect condition except that they seem to be too small for the pulley they're riding on. This happens when the sides of the belt have worn down to make the belt narrower than when it was new.

Belts that squeal on startup or during operation are almost certainly under tensioned. Another sign of under tensioned belts is excessive belt vibration or "slap." Under tensioned belts slip and slipping belts will wear much faster than properly tensioned belts.

Belts which have far exceeded their useful life can begin to separate. The rubber on the bottom (inside) of the belt will begin to break away from the fabric backing leaving large chunks of belt missing. You can usually hear the clunk-clunk-clunk of these belts passing over the pulleys before even opening the machine for inspection. The best way to avoid this is to inspect belts and replace them before they reach the end of their life. Belts in continual use should be replaced annually if not before.

INSPECTING BELT DRIVES

Belt drives that operate continuously should be inspected every 3 months. Since most drive belts are found on air moving equipment

(RTUs, air handlers, etc.), these inspections are usually scheduled to be performed with filter changes or other HVAC inspections.

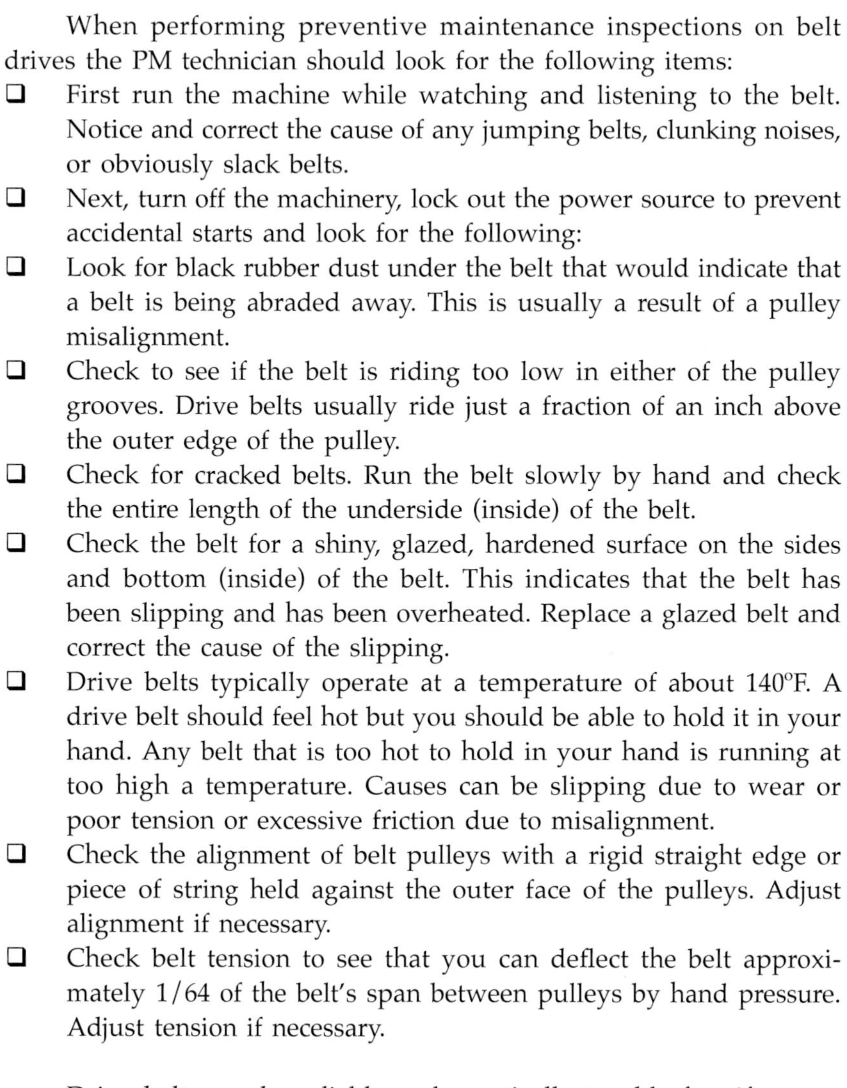

When performing preventive maintenance inspections on belt drives the PM technician should look for the following items:

- ❑ First run the machine while watching and listening to the belt. Notice and correct the cause of any jumping belts, clunking noises, or obviously slack belts.
- ❑ Next, turn off the machinery, lock out the power source to prevent accidental starts and look for the following:
- ❑ Look for black rubber dust under the belt that would indicate that a belt is being abraded away. This is usually a result of a pulley misalignment.
- ❑ Check to see if the belt is riding too low in either of the pulley grooves. Drive belts usually ride just a fraction of an inch above the outer edge of the pulley.
- ❑ Check for cracked belts. Run the belt slowly by hand and check the entire length of the underside (inside) of the belt.
- ❑ Check the belt for a shiny, glazed, hardened surface on the sides and bottom (inside) of the belt. This indicates that the belt has been slipping and has been overheated. Replace a glazed belt and correct the cause of the slipping.
- ❑ Drive belts typically operate at a temperature of about 140°F. A drive belt should feel hot but you should be able to hold it in your hand. Any belt that is too hot to hold in your hand is running at too high a temperature. Causes can be slipping due to wear or poor tension or excessive friction due to misalignment.
- ❑ Check the alignment of belt pulleys with a rigid straight edge or piece of string held against the outer face of the pulleys. Adjust alignment if necessary.
- ❑ Check belt tension to see that you can deflect the belt approximately 1/64 of the belt's span between pulleys by hand pressure. Adjust tension if necessary.

Drive belts can be reliable and practically trouble free if proper installation and maintenance techniques are followed. Poorly installed or maintained drive belts can be a big time waster for the maintenance department. Drive belt failures also have one of the largest impacts on your building's occupants since drive belts are primarily used for comfort control equipment. Belt drives are one of the building components that benefit significantly from a simple but effective PM program.

CHAPTER 8 SUMMARY

- Unsatisfactory installation, improper belt selection, and poor adjustment make belt drives account for a very high percentage of preventable mechanical failures.
- The three different types of drive belts commonly in use are "classic cross section" belts, "narrow profile" belts, and "fractional horsepower" belts.
- Classic cross section belts have cross sections designated as A, B, C, or D.
- Narrow profile belts have cross sections designated as 3V, 5V, and 8V.
- Fractional horsepower belts have cross sections designated as 2L, 3L, 4L, and 5L
- All drive belt part numbers give the belt length in inches if you know how to do the conversion for each type of belt.
- High torque applications make use of multiple drive belts side by side on multi-grooved pulleys. These belts are ordered as matched sets to ensure that the length of each belt in the set is the same.
- A properly tensioned and aligned drive belt should last one year of continual operation.

Chapter 9

Indoor Air Quality

Indoor air quality (IAQ) is a relatively new topic that has recently exploded as a major issue for facilities maintenance departments across the country. IAQ problems in the workplace have been a familiar news story in recent years and concerns among building users is on the rise. Even with widespread IAQ concerns, there seems to be a lot of mystery among facilities managers about IAQ and how to solve IAQ problems. In this section, we'll try to answer many of the mysteries surrounding IAQ and identify some of the preventive maintenance that can prevent IAQ problems from ever starting. Preventive maintenance can have a significant impact on the quality of indoor air. Every maintenance manager and PM technician should be familiar with how their work impacts the indoor environment.

The energy crisis of the 1970s helped to spawn many of the indoor air quality problems in facilities. The energy crisis led commercial buildings to be built to a new standard of energy efficiency. For the first time, architects and contractors were vigilant to prevent the loss of conditioned air to the outside and infiltration of outside air into a facility. Buildings built prior to the energy crisis were not built to such air tight standards. In a time when fuel was cheap and abundant, leaky, energy wasting buildings were perfectly acceptable. Leaky buildings don't buildup concentrations of indoor air pollutants and don't have problems with IAQ. Over the past 3 decades, thousands of new, nearly air tight, facilities have been built and as a result, IAQ problems have become a significant issue for many maintenance departments.

THE 3 PARTS OF IAQ PROBLEMS (THE 3 P'S)

As our first step in understanding IAQ, let's first identify what constitutes an indoor air quality problem. All indoor air quality problems have three components; we'll call them the three Ps. The three P's are:

1. **Pollutant** source
2. **Pathway** for the pollutant to reach people
3. **People** who are affected by the pollutant

The **simple** relationship between the 3 P's is **pollutants** move through pathways to reach **people**.

Pollutants—The First P

The sources of these pollutants are nearly unlimited. Tobacco smoke, pollen, paint fumes, cleaning chemicals, and dust from maintenance activities are just a few of the common pollutant sources. Some pollutants such as sewer gas are perceptible by nearly everyone and nearly everyone would consider the pollutant to be a problem in the workplace. Other pollutants such as pollen or mold spores only present problems to a few individuals and may not even be noticeable by most. Nearly anything can be a pollutant if there is someone present who is sensitive to that particular contaminant.

The Most Common IAQ Pollutants

Biological Contaminants—These include mold, dust mites, pet dander, viruses, bacteria, or droppings from rodents or insects. Most of the biological contaminants are found in damp or wet areas such as air conditioners, bathrooms, kitchens, or in areas that have roof or other leaks. Dust mites, mold, pet dander, and insect or rodent droppings can trigger asthma or allergic reactions in building occupants. Bacterial diseases such as *Legionella* can be associated with moisture in ventilation systems.

Dust—often a result of maintenance activities, poor housekeeping, or dirty and clogged air filters bypassing air around the filter.

Tobacco Smoke—contains more than 40 chemicals know to cause cancer and many others that are irritants. Exposure can trigger asthma in some individuals.

Radon—a colorless, odorless radioactive gas produced by the decay of uranium in rock below the soil. Can enter buildings from the soil and is undetectable without laboratory testing. Prolonged exposure is linked to lung cancer.

Asbestos—a material used in many building materials prior to 1975. Asbestos can be found in floor tiles, insulation, wall board, ceiling tiles, construction adhesives, and other places. Tiny asbestos fibers are associated with asbestosis (hardening of lung tissue) and some cancers. As long as asbestos remains undisturbed and intact it presents no hazard. Once disturbed, airborne asbestos fibers present a serious health risk.

Volatile Organic Compounds (VOCs)—solvents found in paint, stains, caulk, and some industrial cleaners. VOCs in the air typically have a strong odor and can cause many health symptoms in sensitive individuals such as headaches and nausea. Long term exposure is linked to a wide range of mental and physical ailments.

Mold Spores—sources can be mold growth inside a building or spores can be brought into a building from outside through the building's HVAC system. Mold spores exist everywhere and only a few species trigger asthma or other allergic reactions in sensitive individuals. An environmental consultant can have surface dust tested to see what types of mold are present.

Cleaning Products—perfumes or other chemicals found in housekeeping chemicals may cause headaches, nausea, or other allergic reactions in some individuals.

Pesticides—today's pesticides are used in extremely low concentrations and are more target specific than in the past. The inert ingredients in some pesticides may cause eye, ear, or throat irritation and although very rare in commercial buildings, prolonged exposure to high levels of some pesticides can cause nervous system damage, liver damage, and cancer.

Pathways—The Second P

Pollutants are only a problem when they make their way to people. The route they take to get there is the pollutant's pathway. Pathways can be any route from the pollutant source and can be difficult to determine. Pollutants can travel through ductwork, above ceilings, down corridors, through open doorways, or even through cracks and gaps in building construction.

Pollen can be drawn into a building through an air conditioning unit's outside air damper and delivered via ductwork through an entire building; Sewer gas can travel through a dry floor drain trap into a kitchen; or the unpleasant odors associated with a locker room can travel through the open locker room door into a gym.

People—The Third P

Contaminants in the air do not cause a problem unless they bother people. Everyone reacts differently to different contaminants. Those of us who work in facilities are constantly in contact with all sorts of air pollutants. Everyday we're exposed to sawdust, adhesives, solvents, welding gases, paint, sewer gas, and all types of dust and dirt. Most of us aren't very sensitive to these things. If we were sensitive, we would have chosen a different line of work. Because most air contaminants don't bother us, we may tend to underestimate the effects of these things on other people.

Indoor air problems can be subtle. In many cases, one or only a few individuals are affected while no one else shows any symptoms at all. In these cases, it can be easy to dismiss the complaint as unfounded since no one else has complained. Not all suspected IAQ complaints turn out to be valid IAQ problems. However, different individuals react differently to air quality issues and all IAQ concerns deserve to be properly investigated.

Different sensitivities among individuals are one of the things that can make solving IAQ problems so difficult. It can be very difficult to track down the source of something in the air if you aren't able to detect it yourself or when you don't know what type of air-born contaminant is causing a problem

Most of the IAQ issues we deal with on a regular basis are easy. Someone smells garbage, paint, or sewer gas and we can figure out where the odor is coming from and can eliminate it. Most of these IAQ problems are so common and so easy to solve that we never even consider them to be IAQ problems.

Unfortunately, the most often discussed IAQ issues are the ones that are difficult to figure out. Many times, an IAQ concern is just a short list of symptoms that an occupant experiences only while they are in the building. Some of the most common symptoms reported as suspected IAQ problems are:

- Headache
- Fatigue
- Shortness of breath
- Congestion
- Malaise
- Coughing and sneezing
- Dry eyes, nose, throat
- Dry skin or rash
- Shortness of breath
- Nausea

While all of these symptoms can legitimately be caused by IAQ, they can also be caused by dozens of other ailments. Knowing when a physical symptom is actually IAQ related can be difficult or impossible.

COMFORT ISSUES

To confuse matters further, many of the IAQ complaints a maintenance department receives are actually comfort complaints that have nothing to do with contaminants in the air. Comfort issues typically involve temperature or humidity levels and not pollutants.

Temperature and humidity problems, particularly low humidity, can cause many of the same symptoms as IAQ problems. Dry eyes, nose, throat or skin, skin rashes; coughing, fatigue, and headache can be associated with low humidity levels.

The American Society of Heating, Refrigeration, and Air-Conditioning, Engineers (ASHRAE) publishes a standard (ASHRAE standard 55-1992) that describes the temperature and humidity levels that are comfortable for the majority (80%) of people in a building. Figure 9-1 summarizes the standard.

Extremely low humidity can cause respiratory problems unrelated to IAQ. However, both high and low humidity extremes can cause IAQ problems. High relative humidity can cause mold growth while extremely low relative humidity can cause mold spores to be released into the air. Mold is a common allergen and asthma trigger.

Another common issue with occupant comfort is the amount of

Relative Humidity	Winter Temp	Summer Temp
30%	68.5°F - 75.5°F	74.0°F - 80.0°F
30%	68.5°F - 75.5°F	74.0°F - 80.0°F
30%	68.5°F - 75.5°F	74.0°F - 80.0°F
30%	68.5°F - 75.5°F	74.0°F - 80.0°F

Figure 9-1. ASHRAE recommended temperature and humidity ranges for human comfort

fresh air introduced into a building. ASHRE standard 62-1989, *Ventilation Standard for Acceptable Indoor Air Quality*, recommends 15 to 20 cubic feet per minute (CFM) of fresh air be brought into every space for each occupant. That means that if an office has 10 people working in cubicles, requiring 20 CFM each, the total fresh air introduced to that space should be 200 cubic feet every minute. This can be accomplished by bringing 200cfm of fresh air into the room or by bringing in fresh air at another location and exhausting 200 CFM from the room so that fresh air is drawn into the room from other parts of the building.

The 15CFM and 20 CFM figure has changed many times over the years and has been settled on as the amount of fresh air that the majority of people (80%) consider to be comfortable. Below this air exchange rate, some building users will describe the environment as "stuffy" and may complain of human odors while many people will find the space to be "drafty" at air exchange rates above these recommendations.

An exchange rate of 20 CFM per person should keep the carbon dioxide (CO_2) levels at around 1000 parts per million (ppm). As occupants breathe, they use oxygen and exhale carbon dioxide. The carbon dioxide levels will continue to rise if no fresh air is brought into the space. CO_2 levels above 1000 ppm is generally considered to be an indication of poor air exchange. Although, in older buildings built to older standards, properly operating HVAC systems may allow a CO_2 concentration to be slightly above 1000ppm.

Dangerous levels of CO_2 are not likely to be found in facilities. OSHA has established 5000 ppm of CO_2 as the permissible exposure limit for 8 hours of exposure per day and 30,000ppm as a permissible exposure limit for exposures of 15 minutes or less. ASHRAE recommends that indoor CO_2 not be higher than 700 ppm above the level found in the outside air to minimize human odors and maintain comfort. CO_2 in the outdoor environment varies but is typically somewhere around 300 ppm.

SOLVING IAQ PROBLEMS

Just as IAQ problems include the 3 P's, pollutants, pathways, and people, the solution to IAQ problems will be in the areas of the 3 P's. We'll look at each one separately.

Pollutants—Solving Problems

Since all IAQ problems include a pollutant, the simplest solution may be to eliminate the pollutant source. For many pollutants such as cleaning chemicals, paints, and other maintenance chemicals, alternative products can be chosen. Pollutants such as solvents, gasoline, or stored pesticides can be moved to another storage location, preferably outside of the building. Garbage odors or vehicle exhaust can be drawn into buildings through fresh air intakes. These items can often be moved to another location. Environmental tobacco smoke is often a source of complaints that can be solved by implementing a no smoking policy or designating smoking areas away from work areas.

When a pollutant can't be substituted or moved, a pollutant in the air can often be exhausted from a building by proper use of ventilation. Using exhaust fans or even opening a door or window at the pollutant source can prevent the pollutant from entering other parts of the building. Exhaust fans in rest rooms, kitchens, and parking garages are common examples of ventilation at a pollutant source.

Pollutants from outdoors such as pollen or dust can often be filtered from the air entering a building. Filters with MERV (Minimum Efficiency Reporting Value) ratings between 5 and 8 usually provide adequate dust removal for most situations. Removing mold spores usually requires more dense filters with MERV ratings between 9 and 12. Recirculated building air can also be filtered to remove contaminants and prevent relocating contaminants from one location to another. Changing from low density filters to higher density filters may require equipment modifications since denser filters tend to restrict air flow.

Pathways—Solving Problems

When a pollutant source can't be eliminated, it may be possible to eliminate the pathway the pollutant uses to reach people. This can be done in two ways. The first is to provide a physical barrier to air movement and the second is to change pressure relationships to make air flow in a different direction.

Creating a physical barrier to air movement can be as easy as closing a door. Maintenance activities such as painting, gluing, welding, soldering, or sawing can put contaminants into the air. Simply keeping the maintenance department or boiler room doors closed to the rest of the building can eliminate pollutants from getting into the rest of the building.

Temporary physical barriers can be created by hanging plastic sheeting around maintenance work to keep dust contained. This is commonly done to contain asbestos fibers, lead dust, and airborne mold during abatement and remediation work.

The second way to block pollutant pathways is to alter pressure relationships. In a building, air is constantly moving from areas of high air pressure to areas of low pressure. These pressure differentials exist from several sources. Heated air tends to rise through a building creating a low pressure area near the bottom of buildings, a phenomenon known as "stack effect." Mechanical ventilation creates zones of low pressure while make up air fans cause localized areas of high pressure. Wind blowing against open windows and doors can create high or low pressures within buildings depending on the direction of the wind.

Since air always moves from high pressure to low pressure, we want to have our lowest pressures at pollutant sources. By doing this, we are bringing air to the pollutant instead of delivering polluted air to the rest of the building. As already discussed, mechanical ventilation with exhaust fans at the pollutant source is a common solution to problems of contaminated air. While ventilation fans at the pollutant source are exhausting the pollutant to the outside, they are also causing a low pressure zone at the pollutant. This low pressure causes air flow toward the pollutant preventing polluted air from flowing toward building occupants.

People—Solving Problems

When a pollutant can't be eliminated at the source and its pathway can't be blocked, the only solution left is to make changes to the people being affected. These changes usually mean moving those that are sensitive to the pollutant to another location or limiting the time spent in the affected area.

Letting an employee swap offices with someone less sensitive to a particular allergen is a workable solution in many cases. Work spaces can sometimes be rearranged to adapt areas with poor IAQ into storage or other rarely occupied space.

If an individual is having problems such as allergic reactions, headaches or other medical issues, he should see his family physician. More information about solving sick building syndrome problems is discussed below.

The 8 Common IAQ Solutions

1. Physically removing the pollutant source from the building
2. Finding a substitute product for the pollutant
3. Removal of the pollutant by increasing exhaust ventilation at the pollutant source
4. Bringing in fresh air to dilute the pollutant
5. Eliminating a particulate pollutant with proper selection of air filters
6. Creating a physical barrier between the pollutant and affected people
7. Rebalancing pressure relationships with mechanical ventilation
8. Relocating affected people away from the pollutant

THE "M" WORD

The IAQ contaminant with the biggest fear factor is probably mold. Mold is such a public panic issue that it's often referred to as "the M word" among facilities managers.

Molds will be found everywhere in a building. Mold can grow on practically any organic material as long as air and moisture are present. Mold can even thrive on the surface of non-organic materials by surviving on organic surface dust. It would be impossible to eliminate all of the mold and mold spores within a building. If you were to take a dust sample from any location and look at the sample under the microscope, you would find several species of mold.

There are many different types of mold. Some molds can trigger allergic reactions or asthma attacks in individuals sensitive to molds. Some species of molds produce toxins called mycotoxins which can cause a wide range of health symptoms in people, some very serious.

Often when an individual suspects they are suffering from sick building syndrome, they will assume the problem is related to mold. Mold can be a trigger for many health problems. An allergist can determine if a person is sensitive and to which species of mold. An environmental consulting contractor should be able to take dust samples to determine if the types of mold present are the same mold species causing allergic symptoms.

The best way to prevent mold problems is to keep moisture out of your building with proper preventive maintenance of the building envelope. When carpets, drywall, or other building materials get wet, they must be thoroughly dried within 48 hours. Carpet extractors, wet vacuums, drying fans, and portable dehumidifiers can speed the drying process.

Once mold does become established, it must be removed. Mold on the surface of non-porous materials such as stone, laminate, or plastic, can be wiped clean using a 10% solution of chlorine bleach. If mold has grown on porous materials including wall board, ceiling tiles, or carpet, the material must be removed. When removing mold covered material be sure to wear personal protective equipment including goggles, gloves, disposable overalls, and a respirator with HEPA filters.

Mold remediation involving large areas, typically more than 100 square feet, are best handled by experienced mold remediators since the EPA recommends full containment. Full containment includes constructing a containment barrier of double layers of plastic, building a decontamination chamber, and creating two air locks to prevent contaminating any other parts of the building.

DIFFICULT IAQ PROBLEMS—SICK BUILDING SYNDROME

Most of the IAQ problems facility managers see are fairly straight forward. There is a known pollutant, causing an obvious problem. Most of these are nothing more serious than an objectionable odor. These complaints are easy to solve by any of the methods mentioned earlier.

Unfortunately, some IAQ problems are not as simple. When a building user complains of headaches, fatigue, or respiratory problems they believe to be caused by poor indoor air quality, it can be difficult to determine if air quality is the culprit.

The term "sick building syndrome" is used to describe cases where building occupants experience physical symptoms while in a building that disappear when the person is no longer in the building.

While many physical symptoms, such as headache, cough, or itchy skin can legitimately be caused by poor IAQ, they can also be caused by dozens of other ailments or stressors such as building lighting, noise, temperature, humidity, poor ergonomics, or even job stress. Facility managers and maintenance supervisors are not physicians and aren't qualified to diagnose health issues. Solving a complex IAQ problem or investigating a suspected case of sick building syndrome will often require the assistance of experts in other fields. If we follow a systematic approach and know when to rely on specialists, even difficult IAQ problems can be solved to everyone's satisfaction.

Collecting Data

The first step in solving a difficult IAQ problem is to collect as much data about the problem as possible. These data should include both the environmental conditions and the occupant's experiences.

The building occupant having symptoms should be recruited to help with the investigation. The occupant should be asked to keep a daily log of their symptoms on an hourly basis; writing down each hour any observations about their own symptoms and their working environment. Including the occupant in the investigation not only provides the data necessary to hopefully solve the problem but also helps to eliminate the mistrust and suspicion which often accompanies IAQ concerns.

Every hour, the person having symptoms should be asked to record the time; any environmental observations such as "too hot," "too cold," "drafty" etc; and how they are feeling at the time. They should be asked to keep this log for at least one full work week.

While these data are being collected, the facilities or maintenance department should be recording temperature and humidity data at least hourly. The easiest way to do this is to place a small temperature and humidity data recorder in the room and record the temperature and humidity throughout the day for several days or even weeks.

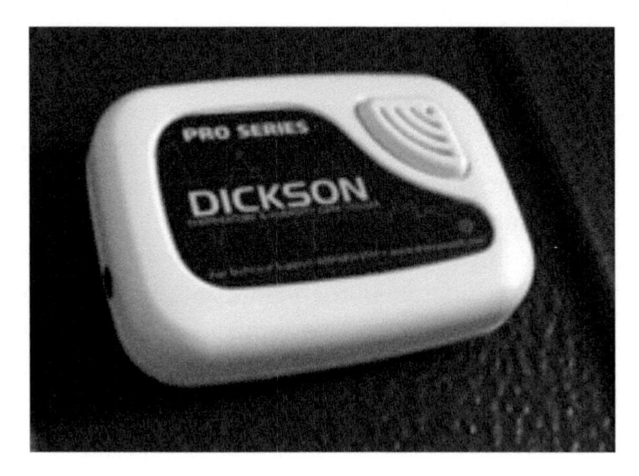

Figure 9-2. Several manufacturers make small temperature and humidity data recorders that can be used for IAQ investigations.

Inexpensive temperature and humidity data recorders are available with prices which start around $100. These tiny recorders can be placed in a location to continually record data and then be taken back to the facility's office to upload the data into a computer. Software provided by the data recorder's manufacturer can chart the data over time.

While temperature and humidity measurements are critical, measuring CO_2 levels can also be helpful in solving IAQ problems. Small inexpensive CO_2 data recorders are available that are nearly identical to the T&H recorders previously mentioned. Typically CO_2 levels under 700 ppm indicate excessive ventilation while CO_2 above 1,200 ppm indicate that ventilation rates are less than desirable.

Data should be recorded for at least a week and the occupant's log of symptoms should be compared to the temperature, humidity, and CO_2 logs that have been collected. A correlation may become obvious. If symptoms of dry eyes, nose, and throat are reported each time the relative humidity dropped below 30%, it should become obvious that the problem is one of low humidity which could be solved by installing or repairing a humidifier. If a staff member feels nauseous whenever the CO_2 level in a room rises above 2000 ppm, there is a strong indication that the space in inadequately ventilated and the stuffy feeling is causing the feeling of nausea.

Seeking Professional Assistance

When IAQ complaints involve human medical symptoms, the maintenance department may not be able to find a cause or solution to anyone's satisfaction. For most of us, our expertise is not in environmental science or medicine. Like all other facilities and maintenance problems, if we don't have the necessary expertise in-house to solve a problem, we hire outside contractors who do. When we are dealing with health issues which may be related to indoor air, the experts we need to rely on are doctors, your local or state department of health, and independent environmental consultants.

Any building occupant complaining of health symptoms should be encouraged to see his or her doctor. If the symptoms seem to be related to an allergen, they will most likely be referred to an allergy specialist. An allergist will likely perform a test to determine which specific allergens cause the patient to have symptoms. Ask the employee if they would be willing to share a copy of these results with you for the purposes of finding a solution to the problem. Also be sure to ask if the employee's physician had any suggestions as to the cause of the symptoms.

Keep in mind that any medical test or treatment is confidential and an employee is under no obligation to provide an employer with this information. If an employee does volunteer to share this information, you cannot share the information with others, such as your environmental consultant, without the employee's permission.

The next step in your investigation will probably be to hire an environmental consultant to help with your investigation. A qualified consultant that specializes in IAQ will have the necessary skills to solve any legitimate IAQ problem. Most of these individuals either have degrees in environmental science or are certified industrial hygienists who specialize in understanding how the workplace environment affects workers.

Your environmental consultant will probably take air or dust spot samples and have these tested at a laboratory to see if they contain any of the allergens that were determined by the effected employee's allergist. If allergens are present, a variety of methods will be used to track down the source of the problem. Air sampling gives us a snapshot of what is in the air right now. Dust samples tell us what pollutants have been in the air over a much longer period of time.

Occasionally, the news media reports on a disease cluster in a particular industry or a particular workplace. A disease cluster is an unusual grouping of a health problem such as a group of individuals from one department getting a particular form of cancer or several women in one building having miscarriages within a short period of time. Some diseases such as *Legionella* or tuberculosis tend to occur in clusters of people who work or live together.

Most suspected cases of disease clusters turn out to be statistically insignificant, meaning that the apparent cluster is simply a coincidence. However, throughout history, many industries had clusters of disease caused by problems with the air quality in the workplace. The most well known of these is probably asbestosis or specific cancers that result from exposure to asbestos fibers. Whenever there is concern there may be a disease cluster in a facility, you should not hesitate to request assistance from your local or state department of health.

EFFECTIVE COMMUNICATION

There may not be any other facilities issue that can escalate from concern to panic as quickly as concerns about IAQ. Maybe it's fear of what can't be seen or fear of possible outcomes such as cancer, asbestosis, or other health risks. Even when the risks are small or when suspected problems are unfounded, panic and mistrust are always lurking around the corner wherever IAQ is concerned.

The rule concerning IAQ investigation should be to act quickly. Some IAQ worries can be dangerous and should be treated as urgently as any life-safety issue would be treated. Acting quickly can protect people from actual harm if it exists and helps assure staff that the organization takes their concerns seriously, and is being diligent in solving the problem. Not responding quickly validates fears that the company knows about a health risk and is hiding something. It's not enough to just investigate quickly. Concerned staff need to be kept informed.

IAQ investigation can be a lengthy process. Time spent collecting data or waiting for an environmental consultant's report can be seen as stalling. It is vital to make an effort to keep staff informed at every step in the process and to regularly remind them that action is being taken to find a solution. To avoid problems of rumors starting from a misunderstood or forgotten conversation, put it in writing. Regular and frequent emails or memos to staff can have a huge impact on how your investigation is received. A brief note explaining what has been done so far to solve the problem and a timeline for work to come will ensure that everyone has accurate information and not harmful rumors. Consider sending these informational emails or memos not only to the staff immediately affected but to the entire department, floor, wing, or the entire facility.

PM TASKS THAT EFFECT IAQ

Many of our normal PM tasks will have an impact on our buildings indoor air quality. Chapter 11 lists many PM tasks that should be performed. A few that effect PM are:

Pouring water into floor drains monthly prevents floor drains from drying out and losing their seal against sewer gasses. Sewer gas entering a building is a very common cause of bad odor IAQ complaints.

Changing air conditioner filters quarterly keeps air flowing through the filter to remove larger pollutants and dirt particles from the air. If we don't change filters, the filters will eventually become so clogged with dirt and dust that air will find a path around the filter and will not be filtered at all.

Having boilers, furnaces and other gas burning appliances serviced annually will help to assure that burners are operating efficiently and not creating poisonous carbon monoxide gas. Changing CO detector batteries twice a year and changing CO detectors every two years also helps to keep dangerous CO away from the third P, people.

Inspecting exhaust fans quarterly and changing belts and greasing bearings will maintain air exchanges within your building at appropriate levels. Maintaining proper air exchange rates prevents the build up of human odors and carbon dioxide. Keeping exhaust fans operating in areas where odors are generated such as rest rooms, kitchens, and maintenance shops will help to stop IAQ odor complaints.

Roof inspections are intended to prevent water from entering a building. Wet building materials can often result in mold growth. Many individuals are sensitive to specific species of molds and can cause allergic reactions or trigger asthma. Roof inspections can therefore have an impact on the indoor environment.

Keeping air conditioner condensate drain lines clear of obstructions will prevent standing water and mold growth in condensate trays. These are only a few of the many PT tasks that can have a positive impact on the quality of indoor air.

CHAPTER 9 SUMMARY

- Indoor air quality problems involve the three P's: pollutants, pathways, and people.
- Issues such as temperature, humidity, and air exchange rate are comfort issues and are not necessarily IAQ issues.
- IAQ problems can be solved in three ways, the three P's.
- Pollutant solutions include physically removing the pollutant source from the building, finding a substitute product for the pollutant, removal of the pollutant by increasing exhaust ventilation at the pollutant source, bringing in fresh air to dilute the pollutant, or eliminating a particulate pollutant with proper selection of air filters.

- Pathway solutions include both creating a physical barrier between the pollutant and affected people and rebalancing pressure relationships with mechanical ventilation.
- People solutions involve relocating affected people away from the pollutant or limiting exposure time.
- The term "sick building syndrome" is used to describe cases where building occupants experience physical symptoms while in a building that disappear when the person is no longer in the building.
- Suspected cases of sick building syndrome require the collection of data, including physical symptoms, temperature, humidity, and carbon dioxide levels.
- If a building occupant suspects that they are suffering symptoms from something at work they should be encouraged to see their physician to help determine the actual cause of the symptoms.
- Environmental consultants can offer valuable assistance in investigating and solving IAQ problems.
- Keeping building occupants informed during IAQ investigations will help to prevent the fears, mistrust, and panic which often accompany IAQ concerns.
- Many of the PM tasks that we perform can have a positive effect on IAQ and can prevent many IQA problems.

Chapter 10

Paint and Protective Coatings

One of the easiest ways to ruin just about anything is to leave it out in the weather. UV rays from the sun, rain, freeze-thaw cycles, and wear from wind-blown dirt all contribute to the failure of anything left outdoors. Of course, buildings are subject to those damaging elements all the time. We can't bring buildings indoors to protect them from the weather. To minimize damage caused by being out in the elements, we put rubber, asphalt, or tar on the roof, and then cover the rest of the building with a protective coating of paint.

Paint makes a significant impact on the long-term condition of our buildings. Left exposed to the weather buildings quickly become damaged. Something as simple as applying a protective coating of paint can prevent almost all of this damage.

However, applying a protective layer of paint is not quite as simple as it seems. Proper selection and application of paint is necessary or the paint will fail. Failed paint can trap water, salt, corrosive pollutants, or other damaging substances in or against the building and can make problems worse than if paint had never been applied. Even if the paint itself does not fail, the wrong coating can allow contaminants and moisture to migrate through the paint causing damage.

A regular schedule of painting interior surfaces should be included as part of a successful PM program. Interior painting is more of an aesthetic issue than a matter of protecting the surfaces. For this reason, we will be limiting the scope of this chapter to exterior coatings which work to protect and extend the life of buildings.

TYPES OF PAINT

The science and chemistry of paint has changed dramatically in the past several decades. Changing environmental regulations since the 1970s and the development of new paint materials including epoxies and, urethane paints, have given the facilities manager new choices in architectural coatings. The vast majority of exterior architectural coatings still fall into the same two general categories which have been available for half a century: latex-based paint and oil-based paint. Since these two classes of paint make up the majority of paint used on facilities and have been protecting buildings successfully for so long, most of our discussion in this chapter will be limited to these two power houses of protective coatings.

These types of paint are made of three distinct components: The vehicle or solvent, the binder, and the coloring pigments. The pigments or coloring agents for both classes of paint can be the same but the solvents and binders differ significantly.

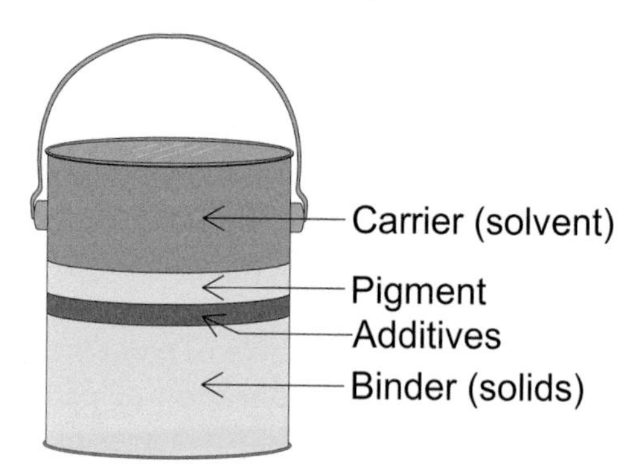

Figure 10-1 Components of paint

The paint vehicle, carrier, or solvent is the part of the paint which evaporates away as the paint dries. In latex paints, the carrier is water with some additives such as glycols or glycol esters. Oil-based paints use petroleum-based solvents, most often mineral spirits. The purpose of the solvent is to keep the binder and pigment suspended in solution in a state that can be brushed or sprayed onto the surface.

The second component of paint is the binder, sometimes referred to as the "solids" of the paint. This is the part of the paint that remains after the paint has dried. The binder, as the name implies, binds the paint to the surface and binds the paint to itself. More solids in a paint results in a thicker film, a more durable finish, and better hiding of undercoats.

The binders used in oil-based paints may be synthetic resins, specifically: alkyds, silicones, and polyurethanes. They can also be vegetable oils including cottonseed, linseed, soybean, safflower, or tung oils. The most common binder in oil-based paints are alkyds, which is why the terms "oil-based" and "alkyd-based" paints are used interchangeably. Alkyds are synthetic resins made by reacting an alcohol with an acid, hence the term "alkyd." Water-soluble binders in latex paint include plastics and polymers such as acrylic resins, polyvinyl acetate, and styrene butadiene. On hundred percent acrylic paints are the most durable latex paints followed by vinyl acrylics and poly vinyl acetates.

The binders in paint go through a chemical bonding process as the paint dries. As alkyd or latex-based paints dry, the small molecules of binder material link together to form large polymer structures. This process makes a very strong final coat of paint because the entire film is one contiguous structure of cross linked molecules.

The third component of paint is the color pigments. There are many organic and inorganic pigments that can be used in both oil and latex paints. When the solvent component of paint evaporates away, the remaining finish film is a combination of the polymerized binders and the entrapped pigments. The higher the concentration pigments, the flatter the paint appears. The lower the concentration of pigments, the glossier the final paint appears. Titanium dioxide, also called titanium white, is a natural inorganic pigment found in most paint formulations. Inorganic pigments tend to have better color retention than organic pigments. Other pigments and small amounts of pigment extenders or other additives are added to alter the color, sheen and wearability as desired for a particular paint.

Oil-based Paint

Oil-based paints are named for the petroleum derived solvents used as the carrier. These paints are also referred to as alkyd paints, a reference to the most common type of binder used in oil-based paint. Oil-based paints make up less than 30% of the architectural paints used today. Oil-based paints were once the most common type of paint, but environmental regulations and advances in water-based paint technology have pushed oil paints into the number two position.

Oil base paints have some advantages compared to latex paints and one type of paint will not be right for every application. Oil-based paints can be applied at lower temperatures than latex paints and still provide a strong, durable coat. Oil-based paints tend to form a harder film than latex paints. This hard film is the reason oil-based paints are sometimes called "enamels," an allusion to the extremely hard vitreous enamel used in plumbing fixtures.

Oil-based paint forms a film that is impervious to moisture migration. This can be a good feature if you wish to keep moisture from migrating into a structure or keep stains from migrating up to the surface of the paint. It can also be a bad feature if moisture from inside a building gets trapped and lifts the paint film causing paint blisters.

Oil-based paint adheres better than latex paint to dirty, glossy, or weathered surfaces and is well suited for application over old oil-based paint with a chalky surface.

Oil-based paints also have several disadvantages over latex paint. Oil-based paints need 6 to 8 hours to dry and need 24 hours between re-coating. The paint will continue to dry indefinitely until, after several years, the paint crystallizes and cracks. Oil-based paint can also only be applied over other oil-based paint, and cannot be applied over latex paint.

Old containers of oil-based paint are considered hazardous waste under federal regulations with considerable disposal costs. Clean up requires solvents such as turpentine, paint thinner, or mineral spirits.

Latex Paint

Latex paint has its own advantages over oil-based coatings and is considered to be a superior product in most, but not all, applications. One hundred percent acrylic latex paints are the best of the latex paints. The most notable difference between oil and latex-based paint is the minimal odor of latex paints compared to oil-based paint. The use of water as the carrier has eliminated most of the indoor air quality (IAQ) and environmental concerns associated with oil-based paint. Latex paint can be disposed of with regular waste (once dried) and does not present any special environmental disposal concerns.

Oil-based paints must be applied to surfaces that are completely free of any moisture. Latex paints can be applied to damp surfaces and surfaces can even be dampened before painting to retard drying in hot weather or when painting in the sun. Latex paints allow moisture to migrate through the paint to prevent most problems with paint blistering. However, allowing water to migrate into the building may not be a desired result in all applications.

Latex paint dries in minutes and can be recoated in only four hours. Latex paint also retains its color better than oil-based paint and resists yellowing better than oil-based paints. With proper surface preparation, latex paint can be applied over oil-based paint but not the other way around. Latex paint is also easier to apply as it is less sticky than oil-based paint and flows and spreads easier and is easily cleaned up with soap and water.

Latex paints have a few disadvantages over oil-based paints that make oil-based paints more suitable for some applications. Because Latex paints allow moisture to migrate through the paint, they are more vulnerable to bleed through of stains. Latex paints provide excellent adhesion to most surfaces but cannot compare to the adhesion of oil-based paints on dirty or chalky surfaces. Latex paints also have a finish coat that is somewhat softer than oil-based paints making oil-based paint a good choice for heavy traffic areas subject to abrasive wear such as hand rails.

Elastomeric Wall Coatings

Usually referred to as EWCs, elastomeric wall coatings are flexible acrylic latex paints developed to be applied in very thick films to masonry surfaces such as concrete, stucco, or concrete block. Typical film thickness is roughly ten times the thickness of regular paint making these paints considerably more expensive than other types of paint due to the amount of paint required. Typical EWC application rates are 50 to 60 square feet per gallon with two coats recommended. Elastomeric wall coatings form a strong flexible membrane that can bridge gaps and flex as the surface below moves and cracks expand and contract.

Surface preparation for EWCs require caulking all cracks larger than 1/16 inch with an acrylic or siliconized (not silicone) caulk. If the surface is particularly porous, a latex primer or masonry sealer should be applied first. Elastomeric coatings cannot be applied where there is surface moisture since this moisture will be permanently trapped and will cause blistering of the paint. Water cannot be allowed to get behind EWCs so window caulking, chimney caps, and roof membranes must be maintained in good condition. EWCs should not be used in locations where water can get behind the masonry such as retaining walls. Elastomeric coatings provide excellent protection of masonry surfaces in areas subject to significant amounts of wind driven rain as long as water is prevented from getting behind the finish.

The special formulation of these paints is prone to excessive chalking as weathering causes paint pigments to form on the surface.

Therefore, these paints are usually only available in light colors which require less color pigment than darker colors. If dark colors are desired, a coat of regular acrylic latex paint can be applied over EWC coatings.

Epoxy Paints

Epoxies, or polyepoxides, are formed when an epoxy resin is mixed with a catalyzing agent. The resulting material is hard, durable, and plastic-like. Epoxy paints usually consist of two components, the resin and the hardener, which are site mixed just before they are applied. Less common are water-based, two-part epoxy paints and one component epoxies that cure with exposure to the air. Epoxy paints and coatings are commonly used as a finish floor material over bare concrete, for use underwater, or used to repair rust or corrosion damage to water tanks, cooling towers, or condensate trays.

Epoxy paints are generally limited to indoor use because they maintain color fastness and weather poorly when exposed to the elements.

Urethane Paints

Urethane paints and finishes are a class of materials more similar to epoxies than traditional paints. Two component urethane paints offer superior hardness and strength in certain applications and are more suited to outdoor use than epoxy paints. The terms urethane or polyurethane refer to the binders used in some coatings or clear finishes. Urethanes are created by the chemical reaction of polyisocyanate molecules with another molecule, often a hydroxyl group molecule.

Paints that are marketed as urethane paint are usually a blend of acrylic resin and water-based urethane. The addition of urethane binders makes these paints exceptionally sticky, provides superior adhesion, and creates a truly high performance paint. Six different classes of urethane coatings are manufactured but storage, mixing, and application requirements are difficult enough to limit most of them to industrial use in manufacturing. These industrial urethanes can be heat cure, moisture cure, or can cure by the addition of a catalyst.

The two types of true urethane coatings that are available for architectural use are Type I and Type VI urethanes. Type I is the familiar clear polyurethane varnish available at any hardware store for finishing bare wood. This finish consists of an alkyd binder reacted with polyisocyanate (urethane) to improve the toughness and abrasive resistance beyond that of regular alkyd paints. Type VI urethane paint is a one part coating in which the polyisocyanate is unreacted and suspended in a solvent. A polyisocyanate film is deposited on the surface when the solvent evaporates.

Rust Inhibitive Paints

These specialty paints include pigments that prevent corrosion of iron and steel. Metal pigments, such as zinc, chromate, and lead work by combining with oxygen to make the oxygen unavailable to oxidize the iron in the protected metal. Oil-based rust inhibitive primers do a better job than latex primers unless the paint is being applied over an oil-based prime coat. For rust inhibitive paints to work, they cannot be applied over other paints. The rust inhibiting pigments must be in direct contact with the metal.

In coastal areas, dissolved salt can become trapped under layers of paint where it will quickly cause significant amounts of rust. The same is true in areas that use salt to melt ice in parking lots and sidewalks. In the Snow Belt, it's common to see bubbling and flaking paint on the bottom few inches of doors and door frames caused by road salt corrosion under the paint. Salt resistant paints are available for these special applications. If painting over existing rust, consider using a rust converter discussed later in this chapter prior to repainting.

Mildew-resistant Paints

Mildew-resistant paints come in many different forms. Some mildew resistant paints have a mildewcide added to the formulation. Packets of powdered mildewcide are available from paint suppliers for field mixing with most latex or oil-based paints. Other mildew-resistant paints use zinc oxide as one of the pigments instead of or in addition to the more common titanium dioxide (also called titanium white). Zinc oxide offers some mildew resistance when added to paints.

Latex paints or oil base paints with vegetable oil carriers such as linseed are the most susceptible to mildew with petroleum-based paints offering the best resistance. Gloss finishes are also less likely than flat finishes to support mildew growth.

Cold Galvanizing Compounds

Cold galvanizing compounds are zinc paints used to protect iron and steel from rust. These coatings are available is a brushable compound or in aerosol spray cans for protecting welds, abrasive wear area, or other small items from rusting. Cold galvanizing compounds consists of fine zinc dust mixed with small amounts of epoxy binders in petroleum or organic solvents. When the solvent evaporates, it leaves a film deposit of up to 95% zinc metal on the surface similar to hot dip galvanized protective coatings. Many manufacturers of cold galvanizing compounds claim coating durability and corrosion protection as good or better than true hot dip galvanizing.

PRIMERS AND SEALERS

Primers and sealers are specially formulated paints used as an under coat to improve adhesion, prevent bleed through, protect the final coat of paint against moisture or alkalinity, or to ensure an even appearance of the top coat. Sealers and primers are similar in composition with the primary difference being the amount of pigment added to the container. Sealers are often clear with few or no pigments added while primers have pigments included and produce an opaque undercoat.

Sealers are most commonly used on masonry surfaces such as block, concrete, or stucco, to increase adhesion of the topcoat and to protect a top coat of paint from efflorescence and alkalinity. Clear sealers can also be used alone to prevent moisture from entering masonry surfaces.

One of the functions of a primer is to improve adhesion between the top coat and the substrate or between the top coat and older layers of paint. Primers are especially good at bonding to glossy or dirty surfaces. Lightly sanding of shiny surfaces will help primers to bond more readily. Primer-sealers also protect a top coat from bleed through of stains from below.

Primers are engineered with different goals than the engineering of finish paints. Finish paints are formulated to be hard, durable, and to resist dirt and environmental contaminants. Primers are formulated to adhere and seal well. Primers do not have the weatherability of paints and must be protected with a quality paint topcoat.

Primers are not required in all cases. If you are painting over a layer of existing paint in good condition that has been lightly sanded or deglossed, and is free of dirt and chalking, you will probably not need a primer. If the surface is porous, previously unpainted, dirty, chalky, or noticeably worn, a primer will help to seal the imperfections and help the new paint to bond. While surface preparation is important for any paint job, using a primer can greatly reduce the time spent on surface prep.

Water- and Oil-based Primers

Like paints, there are two broad classes of primers, oil- (or alkyd-) based primers and water- (or latex-) based. Shellac-based primer-sealers that use alcohol as the carrier or solvent are also common.

Water-based primers can perform as well as the more traditional oil-based primers in most applications. Water-based primers offer the same advantages as water-based paints, notably: low odor and easy cleanup. Oil-based primers offer better adhesion to chalking surfaces and are better at preventing stains from migrating through the new layer of paint. If the surface is free of oil or other stains that will bleed readily, latex primers should be adequate.

Shellac Primers

Usually marked as stain-blocking sealers or "stain killers," shellac-based primers are excellent performers on most surfaces. Shellacs bond well to glossy surfaces and do an excellent job of sealing many of the most difficult stains such as, wood tannins, cigarette smoke, water stains, creosote, crayon, and oils. Stain blocking primers are also available in aerosol cans for spot treating small stains prior to painting.

Shellac-based primers are often used to seal interior surfaces to eliminate pet, smoke, or other odors from migrating through walls and ceilings and are a normal part of fire restoration work. These primers are thinner than other types of paints or primers and do an excellent job of penetrating into bare wood, or other porous surfaces.

As good as shellac primers are at creating a bond to difficult surfaces and at sealing difficult stains, they are not well suited to exterior painting. Their use out doors is generally limited to spot sealing of difficult stains prior to application of a suitable exterior oil or latex primer.

Rust-inhibitive Primers

Like rust-inhibitive paints, rust-inhibitive primers rely on metal pigments such as zinc, chromate, and lead to inhibit the oxidation of steel. Oil-based primers perform better in this application because oil-based primers form a barrier to moisture while latex-based primers allow moisture to migrate through the primer.

Bonding Primers

Bonding primers are engineered to bond to very shiny surfaces such as ceramic tile, glass, and laminate. Bonding primers aren't flexible enough to be used outdoors where substrates are constantly expanding, contracting, and flexing.

THE RIGHT PAINT OR PRIMER FOR THE JOB

The information above should help you in deciding which paint or primer is best for your painting projects. Knowledgeable paint sales people will also be able to offer valuable advice at your local paint supplier. A quick guide to paint selection follows:

New Unpainted Wood—Prime with a quality latex or oil-based exterior wood primer. Oil-based primers are better for staining woods like cedar or wood with dark knots.

Weathered Unpainted Wood—Sand the surface to remove all loose and weathered fibers, treat as with new unpainted wood.

Previously Painted Wood—Primer is not necessary if existing paint is intact and free of chalking. If bare wood or chalking is present, use an oil- or latex-based primer. Chalking surfaces should be painted with oil base materials.

Stucco and Other Masonry—Apply a masonry primer or sealer. Masonry primers are formulated to protect the finish coat of paint from efflorescence and alkalinity in the masonry. If repainting over exiting paint, apply a primer if the undercoat is chalky or in poor condition. An oil-based primer is best over old chalky paint.

Ferrous Metals—Remove as much old rust as possible by wire brushing, rinse or wipe off all dust and allow the material to dry. Apply oil-based rust-inhibitive primer before painting. Consider using a rust converter to rusted material followed by an oil-based primer. Rust converters are discussed later in this chapter.

Aluminum, Galvanized Iron—Clean the surface of any oils, white powdery oxides should be removed with a wire brush or steel wool. Apply metal primer to all bare metal. Aluminum is especially difficult to paint. Bare aluminum quickly forms a layer of aluminum oxide on its surface when exposed to the air. This thin layer does an excellent job of protecting the aluminum from corrosion. However, this layer of aluminum oxide also resists the adhesion of paint. Because paint on aluminum is likely to fail so quickly, the best course of action is to leave aluminum bare.

Slick, Glossy Surfaces—Use a specialty bonding primer on glass, ceramic tile, and plastic laminates. For maximum adhesion, sand the surface first with fine sandpaper.

Leaching Stains—Stains of oil, grease, crayon, ink, water stains, and smoke should be spot sealed with a shellac-based stain sealer. An oil-based primer should be applied over the shellac in exterior applications.

PREPARING THE SURFACE

No part of the paint job is as important as getting the surface ready for paint. Preparing the surface can be time consuming, often requiring more work and time than spent applying paint. Doing a good job of surface prep is fairly straightforward.

For all types of surfaces, preparing the surface involves removing any loose, flaking, or corroded material, cleaning to remove dust and dirt, and de-glossing shiny surfaces areas if necessary. If the surface has bare spots, is porous, or can't be thoroughly cleaned, a primer should be applied to help the finish coat bond properly. Once completed, the prepared surface should be clean, and free of loose material including dirt, dust, or chalking.

The following six steps should be followed when preparing a surface for paint. Depending on the condition and material, not all six steps will be required for every paint job but this list will be a useful tool to ensure that no step is missed.

NOTE ABOUT LEAD PAINT: Lead was a common pigment used in paint prior to 1978. Any buildings built prior to this date may contain lead paint. Sanding, scraping, or wire brushing lead paint can release toxic lead dust. The concerns with lead paint are discussed in more detail later in this chapter.

Six Steps to Proper Surface Preparation

Step One

Kill any mildew. Mildew will be found in areas where sunlight rarely reaches such as the north side of a building, under trim, or in areas with dense cover of vegetation. If you apply a coat of paint over a mildewed surface, the mildew will grow through the new layer of paint. Mildew can't be washed from the surface because it would be impossible to remove every microscopic fungal spore and any spores left on the surface would grow a new outbreak.

Mildew must be killed to prevent regrowth into a new coat of paint. Mildew can be killed with a 25% solution of household bleach sprayed or sponged onto the old paint and left to sit for at least twenty minutes. If the solution dries in this time, it should be reapplied. After the twenty minutes (or longer) rinse the surface with clean water.

Step Two

Remove all loose paint. Scraping, pressure washing, or brushing with a steel brush will remove most loose paint. If you need to remove paint completely, heat guns or chemical strippers can be used. When using chemical strippers, be sure to follow the manufacturer's directions for surface clean up and disposal of the old paint. After stripping, the surface should be sanded to smooth out any rough edges or paint chips. Use wire brushes or steel wool carefully, any metal pieces left on the surface can oxidize creating brown spots in the new paint.

Step Three

Lightly sand any glossy surfaces. A quick sanding with a fine 180 to 220 grit sand paper will roughen shiny surfaces enough to allow the new coat of paint to adhere well. Remove any dust with a damp rag and allow the surface to dry.

Step Four

Caulk. Any cracks, gaps, or penetrations should be caulked. When it's time to repaint, it's probably also time to caulk. Use a quality acrylic or siliconized acrylic caulk. 100% silicone caulks have exceptional adhesion and durability and will last a long time outdoors but will not take paint so avoid silicone caulks in any areas that will be painted.

Step Five

Wash the surface. Wash the surface with a mild detergent and water or if extremely dirty, by pressure washing. A film of dust and dirt on the surface will prevent paint from making a good bond with the surface being painted. Pressure washing can drive water deep into wood or masonry surfaces and ample time, up to several days, should be given for through drying before applying paint.

Step Six

Coat with a good quality primer. Primers are necessary for bare wood, porous surfaces, whenever the old layer of paint is in poor condition, and for difficult to paint surfaces. See the section on primer selection earlier in this chapter for guidance in choosing the right primer for your paint job.

Rust Converters

Rusting iron and steel can present a difficult painting problem. Once rust starts, it can be difficult to stop. Even if you sand or wire brush rusty surfaces to remove loose rust, paint can have a difficult time adhering well. Or worse, a layer of paint over a rusty spot can trap moisture accelerating the corrosion below.

There are products that can work wonders on rusty areas to convert rust to a stronger, more durable compound and to prevent further rust damage to the area. These products are rust converters. Available as a liquid, thicker than most paints, rust converters are applied directly to the rust after scraping or brushing to remove any loose material. When using rust converters, you do not need to remove all of the rust to expose the bare metal, in fact, rust converters do nothing to bare metal and only work if rust is present.

Rust converters contain tannic acid, a naturally occurring weak acid found in plants. The tannic acid reacts with iron oxide (rust) and converts it to iron tannate, a stable black material. When rust forms, it flakes away from the original metal exposing more metal surface to the environment allowing more rust to occur. By contrast, iron tannate bonds tightly to the surface metal preventing any further corrosion.

Rust spots can be treated with a rust converter prior to priming and painting. Loose rust should be removed to ensure that the chemical converter can reach the rust right at the metal surface. If rusting is extensive, a second application will assure complete coverage of any rusted areas. Rust converters dry to the touch in a few hours but should be left alone for 24 hours to allow the conversion process to take place. The surface must be rinsed thoroughly before priming to remove any acid residue. Only oil-based primers should be used over metal treated with a rust converter.

ENVIRONMENTAL CONCERNS

Volatile Organic Compounds (VOCs)

Possibly the biggest driving force today causing changes in the painting and coatings industry is concerns over the release of volatile organic compounds. VOCs are organic chemicals such as the solvents found in paints that vaporize at normal temperatures. VOCs can cause several health problems such as allergic reactions, nausea, headaches, kidney, liver, and central nervous system damage, and many are believed to be carcinogens. Volatile Organic Compounds are one of the most common pollutants contributing to problems with indoor air quality (IAQ). When painting indoors, adequate ventilation and careful selection of paints will help to alleviate most IAQ problems. See Chapter 9 for more information on VOCs and IAQ.

Many VOCs also contribute to environmental problems. Some VOCs react with other chemicals in the air to create ground level ozone and the related respiratory problems for some people. Some VOCs are also greenhouse gasses. The concentration of these compounds was first regulated by the US EPA in 1970 with the passing of the Clean Air Act. Federal Regulation 40 CFR 59.400-413 is National Volatile Organic Compound Emission Standards for Architectural Coatings which covers paints, stains, concrete sealers and other architectural coatings. This regulation and many diverse state regulations and voluntary standards adopted around the country limit the types and amounts of Volatile Organic Compounds that can be included in paints.

It was this regulation on the amounts of VOC solvents in paints that lead to the increased use of water-based latex paint from less than 30% of paint used in the 1970s to more than 80% of paint used today.

The formulation of paints have changed considerably since the 1970s giving the consumer many more choices and making selecting the right paint more complex than ever before.

Lead Paint

Prior to 1978, lead was used as a pigment in many formulations of paint. Lead pigments offered corrosion resistant properties to ferrous materials and had excellent adhesive properties. Lead is also toxic, especially to children. Lead exposure can cause developmental problems, hearing loss, kidney damage, and nervous system damage in children and adults. In 1978, the US Consumer Product Safety Commission banned paint containing anything more than trace amounts of lead for use in residential buildings or buildings occupied by children. Essentially all facilities residential or otherwise stopped using lead paint from that time forward.

Any building built before 1978 may have been painted with lead paint and could still have layers of paint containing lead. Unless you are fortunate enough to be able to apply a new coat of paint without disturbing the existing paint, any sanding, wire brushing, scraping, or washing could release lead dust. Many states require any disturbance of lead paint to be done by accredited lead abatement contractors and require any material removed to be handled as hazardous waste.

Your environmental services contractor will be able to test the paint on your facility and tell you if lead is present before you accidentally cause a hazardous condition for your employees and the public using your facilities.

Paint Disposal

Unused paint presents a different disposal problem. The best solution to disposal of unused paint is to minimize the amount of paint left over after a project. Careful measurement and careful ordering of paint will help to minimize both the cost of the project and the amount of wasted paint requiring disposal. As a commercial customer, you probably have some buying power with your paint supplier and may be able to negotiate taking back unopened cans prior to your purchase, especially if you are using primers, whites, or stock colors.

Unused paint can also be stored for later use. Some paint should be saved for paint touch up later since companies are constantly changing paint formulations, you may have a hard time matching a particular finish later. However, some paints contain chemicals, mostly solvents, which can cause health problems or need to be handled as flammable liquids. Excessive storage of these paints is not a good idea.

Unused latex paint can be disposed of with regular waste as long as the paint is dried. A half inch of latex paint in the bottom of a can will dry quickly and can be disposed of with the paint can. A large amount of left over paint can be dried quickly by mixing in a cardboard box with sawdust or shredded newspaper. Maintenance shops often have sawdust that needs disposal and your organizations business office will probably have bags of shredded paper to contribute. Powdered paint hardeners are available from your paint supplier and when added to left over paint, causes it to solidify.

Unused oil-based paint needs to be disposed of as hazardous waste. Pigments can include heavy metals or other toxins and the solvents include VOCs and are flammable. Your regular waste hauler can probably handle this disposal for you at a cost. If you have large quantities that need disposal, there are companies that specialize in

Figure 10-2. Storing excessive amounts of old paint can be hazardous.

inventorying, packaging, disposal, and recycling of hazardous materials. Most localities have "home hazardous waste days" when residents can bring their hazardous household waste to town hall or the public works department for free disposal. A few locations open these days up to commercial enterprises if they are only bringing small quantities.

PAINT FAILURES

Many things can contribute to paint failures. Failures are caused by poor surface preparation, poor selection of materials, poor application methods, or allowing moisture to get behind the paint. The most common causes of paint failure are outlined here with their respective causes.

Application Temperature

We know that most paint failures can be prevented by proper surface preparation and by choosing the right primer and paint for the particular application. Another factor affecting the success of a paint job is the temperature of the air and substrate during and soon after application.

Paint should not be applied when the temperature is too cold or too hot. The manufacturer's label instructions should be followed and will typically recommend not applying paint at temperatures below 50°F or above 90°F.

Latex paint is more prone to problems from cold applications than oil-based or alkyd paints. As latex paints cure at low temperatures, the binder material molecules cannot link together as readily to form the strong polymer structure desired in the finish coat of paint. The resulting film of paint is easily cracked. At very low temperatures, the paint can dry with tiny surface cracks that effect the paint luster making the paint appear flat or dull.

At low temperatures, paint also takes longer to dry. One of the other common problems encountered with slow drying paint is that wind blown dirt and debris often end up damaging the paint job. There is always the possibility of this happening to fresh paint but the risk is increased with a longer drying time. Paint that has remained wet for a long period of time can have a dull appearance from dust embedded into the finish coat.

Painting at high temperatures can also be damaging to the paint being applied. The polymerization process that alkyd- and latex-based paints go through as the paint cures takes time. If the paint dries too quickly due to high temperature, polymerization may not occur resulting in the same weak topcoat encountered with cold weather painting. Applying paint in direct sunlight or on windy days can have the same drying effect as painting when temperatures are too high. Pre-dampening the surface when applying latex paint can minimize both the surface temperature and the tendency of latex paint to dry too quickly.

Keep in mind the application temperature applies to both the temperature of the air and the temperature of the substrate. The proper application temperature needs to be maintained not only during paint application but until the paint has substantially dried, usually 36 to 48 hours after the paint application.

Common Modes of Paint Failure

Alligatoring: caused when finish coats are applied over undercoats or primers before they are dry. The resulting shrinkage of both the top and under coats causes a pattern of cracks in the top coat with an appearance similar to the skin of an alligator.

Bleed through: Knots in wood, asphalts, grease and oils, creosote, crayon, water stains, or any oil soluble stains or inks can bleed through a new coat of paint. The surface should be cleaned before painting to remove as much of the offending stain as possible and apply a seal coat of shellac-based primer, or oil-based primer sealer. Shellac-based primers are some of the best stain killers but do not survive well on the exterior of a building and should only be used for spot sealing stains and must be overcoated with an appropriate primer.

Blistering: Blistering paint is caused by moisture under the paint lifting the paint from the substrate. Moisture can come from leaks in the building roof or walls, can enter through failing paint, or can be present in wood or masonry before painting. Humidity inside a building can also migrate through the building envelope causing paint to blister

Chalking: Caused by a loose, powdery coating of paint pigment left on the surface of paint as the paint weathers. A normal condition that occurs as paint ages. Chalking surfaces must be sealed with a primer prior to application of a finish coat. Oil primers offer superior adhesion to old chalking paint.

Chipping (Flaking): happens when the paint completely separates from the surface. Usually caused by poor surface preparation. Cleaning and sanding glossy surfaces will prevent this condition

Creeping: Creeping happens when paint beads up on the surface. This is more likely to occur on dry days when painting over a glossy finish. Most paints adhere poorly to glossy undercoats. Lightly sanding the surface to remove the shine will prevent creeping. Creeping can also be caused by applying paint that is incompatible with the paint used in the undercoat.

Efflorescence: Efflorescence is a white salt-like deposit that forms on the surface of masonry. As water migrates through masonry, it carries with it soluble salts which are deposited on the surface when the water evaporates. Efflorescence must be removed before painting or poor adhesion will result. Wire brushing, pressure washing, sand blasting, or washing with a solution of 25% muriatic acid (hydrochloric acid) will remove efflorescence. Muriatic acid is extremely caustic and protective gear must be worn when using it.

Holidays: Areas where paint coverage is poor leaving the undercoat visible through the new paint is often referred to as holidays in the paint. Holidays can be caused by choosing a low quality paint with few paint solids or by not mixing paint thoroughly.

Loss of Gloss: Loss of sheen or "flattening" can have several causes. Poor surface preparation, painting in cold weather (under 50°F), or too much thinning of paint can cause flattening.

Mildew: If not properly treated before painting, mildew on old paint will continue to grow on the surface of new paint. To properly treat mildew, first the source of moisture must be corrected. If mildew is growing on the exterior surfaces, consider cutting vegetation to allow sunlight to reach the area. Mildewed surfaces should be treated with a 25% solution of sodium hypochlorite household bleach left on for twenty minutes before rinsing. Mildew resistant paints offer some protection against new mildew growth. However, if moisture issues are not addressed, mildew will eventually return.

Spalling: Spalling happens when chips or flakes of masonry surfaces break away. Spalling happens when tiny cracks and pores in the masonry surface are not completely encapsulated by the layer of paint. Moisture gets into these small cracks and expands when it freezes making the small cracks larger. After several freeze-thaw cycles, the masonry has cracked enough that chips begin to break free.

Tackiness or Slow Drying: Slow drying can be caused by low drying temperatures, applying too thick a paint coat (drips will be evident), by improper ventilation, or drying in areas with high humidity. In some cases, "salamander" type direct vent heaters are used to warm a space in an attempt to dry paint faster. However, the combustion products are primarily made up of water vapor which increases humidity and drying time. Oils, solvents, or soaps on the surface prior to painting can leach into a new coat of paint causing permanent tackiness.

Wrinkling: If paint is applied too heavy, the top surface of the paint layer will dry and shrink while the bottom surface remains wet. This causes the surface to become wrinkled. Applying paint at the appropriate rate will prevent this from happening. If it has happened, the only solution is to sand the surface smooth and paint again.

Yellowing: Yellowing is caused by sunlight, heat, dirt, or chemical fumes. The only solution is choosing paint appropriate for the particular environmental conditions.

INSPECTING AND RECOATING

In addition to the poor protection a bad coat of paint provides to a building, cracked, pealing, or dirty paint give a facility a run down and unkempt appearance.

Like all aspects of a good PM program, paint needs regular inspections to find small problems before damage occurs to the material below. When inspecting paint, look for cracking, weathering, chalking, rust spots or bleed-though from nail heads. Flaking, cracked, or peeling paint is an obvious sign that repainting is necessary. If the apparent color of paint changes when dampened with a sponge or damp rag, this is evidence that the protective finish of the paint is gone.

It is important to schedule repainting as soon as the existing paint is starting to show signs of deterioration. By recoating a building with a new layer of paint before the old layer completely fails, damage to the substrate can be prevented and an enormous amount of time can be saved preparing the surface for a new coat of paint. Typically, exterior painting needs to be recoated every three to five years.

Figure 10-3. Failing to inspect paint allowed rust to eat through the metal skin of this building. A few dollars spent on paint would have prevented the expense of installing new siding.

CHAPTER 10 SUMMARY

- Although there are several specialty paints available, water-based latex paint and oil-based paints are the two paints used in almost all cases in architectural coatings.
- Paints are made of three components. The binder, which is the solid part of paint that becomes the hardened film, the pigments which give paint its color, and the solvent which evaporates away leaving a film of binder and pigment.
- Neither oil base nor latex paints are the correct choice for all applications. Careful consideration of the characteristics of each must be considered when choosing a paint.
- Specialty paints used on facilities include elastomeric wall coatings (EWCs), epoxy paints, urethane paints, rust inhibitive paints, mildew resistant paints, and cold galvanizing compounds.
- Primers and sealers are applied under paint to increase adhesion of the paint, prevent bleed through of stains, improve the appearance of the top coat, and protect paint from moisture and alkalinity.
- Oil-based, latex, and shellac primers are commonly used and, like paints, must be chosen based on the application surface.
- Surface preparation is necessary for a successful paint job. Removing loose paint, cleaning dirt, and lightly sanding glossy surfaces will ensure a good bond between the surface and the paint.

- Environmental and public health concerns have driven significant advances in paint technology over the past few decades.
- Volatile organic compounds (VOCs) are solvents that readily vaporize and can cause health problems for some individuals and may contribute to greenhouse gasses. These compounds are regulated by the US EPA and many states.
- Lead, a toxic substance, was a common component of paint prior to 1978. Older buildings may still have layers of lead paint. Stripping, sanding, scraping, or brushing lead paint can release toxic lead dust. Special precautions are required when working with lead paint.
- As with any part of a PM program, a regular program of inspections to paint will prevent unseen failures from causing damage to your building.

Part III

Specific Maintenance Procedures and Requirements

Chapter 11

Specific Maintenance Procedures and Requirements

We know why PM is important to both the maintenance department and to the bottom line. We know some of the science and engineering principles behind preventive maintenance tasks. We have also learned several methods of keeping track of all of the scheduled PM, and how to get your staff prepared for your new PM program. You now know more about preventive maintenance programs than most maintenance supervisors or facilities managers. So what do we do with this newfound knowledge? We finally get to stop talking about it and actually do it.

This final chapter is about the actual work that PM requires. This chapter is about the PM tasks that belong on your PM calendar and the individual preventive maintenance jobs that you and your maintenance staff will be performing. Every piece of equipment, building component, or mechanical system listed on the following pages will benefit from a scheduled program of maintenance.

On the following pages, you will find preventive maintenance information for many types of equipment found in facilities. Each piece of equipment is listed in several ways. To help set up your PM calendar, each piece of equipment is listed according to how often it needs PM. In order to familiarize you with PM tasks that are required by regulations, codes, and standards there is a listing of exactly which codes and standards govern specific PM tasks. And finally, there is an alphabetical listing of the equipment ordinarily included in most facilities PM programs with general maintenance requirements for each item.

It is important to note that the PM tasks listed here are general guidelines. Specific manufacturers will likely have specific PM requirements that may vary from the ones in this book. In all cases, the manufacture knows their equipment better than anyone else and their maintenance recommendations should be followed whenever available.

Most common types of equipment are listed in this chapter. You can use this list as a checklist of what to include in your PM program. However, every facility is unique and you may have a few unusual kinds of equipment in your equipment inventory that do not appear here. This list is comprehensive but not complete.

Now that you know all you need to start your PM program, it is time to get going. Your PM calendar is probably not perfect yet. You probably have not had the opportunity to complete all of the training you would like your staff to have, and you have not been able to get all of the new tools and equipment you believe will make your PM program perfect. Do not let any of that stop you. If you wait until the perfect time, you will never start. Do not let the fact that you only have a good PM program, and not a great one, stop you from jumping in with both feet. This brings us to truism #9, and one of the reasons so many PM programs (and other good ideas) never get off the ground.

PREVENTIVE MAINTENANE TRUISM #9
Good is the Evil Enemy of Great

If you wait until your PM program is great, you will probably never start. It is time to put your wrenches and grease guns to work and start maintaining the equipment you have been fixing for so long.

Over the next few months, you will continue to make adjustments and changes until you find exactly what works for your facility. You will figure out what types of equipment will make PM easier for you and what will not. You may even start with a simple paperwork system and decide that your PM program is extensive and successful enough to warrant upgrading to a computerized maintenance management system. Regardless of the current state of your fledgling PM program, the time to start is now.

The information that follows is in three separate lists. Each list includes the same types of building components and equipment organized in three different ways. The first list organizes equipment according to the frequency with which PM is required. The second lists equipment that is required to maintained according to codes or regulations and gives the code or regulation that applies. The third section is an alphabetical listing of roughly 100 pieces of the most common equipment to include in a PM program. This third list suggests PM procedures and technical information about each type of equipment.

LIST 1: PM TASKS LISTED BY FREQUENCY

Please see the alphabetical listing of equipment later in this chapter for preventive maintenance procedures and technical details about each type of equipment.

Daily

Carpets—vacuum high traffic areas
Chillers—cursory inspection visually checking status screen
Cooling towers—cursory inspection
Flag poles
Resilient floors—mop
Safety inspection of swimming pools

Weekly

Carpets—vacuum
Chillers—in-house operating observation
Cooling towers—operate fans, pump, observe operation
Emergency generator—running test
Fire sprinkler—check pressure gauges, inspect control valves (without tamper switches)
Lawn Irrigation—walk lawns during operating season

Monthly

Air compressors
Boiler/water heater inspection (in house)
Carbon monoxide detectors
Circulator pumps
Door inspections and adjustments (entrance doors and smoke and fire doors)
Elevator—in house inspection
Emergency generator—full load test
Emergency lighting and exit signs
Floor drains
Fire alarm in-house test
Fire drill (varies with location and type of facility
Fire extinguisher inspections (visual, in-house)
Fire sprinkler—check pressure gauges, inspect control valves (with tamper switches)

Housekeeping equipment
Ice machines
Kitchen fire suppression system in-house inspection
Laundry (commercial washers and dryers)
Lighting (exterior)
Lighting (interior)
Parking lots
Resilient floors—spray buff or clean with pad and apply additional coats of finish
Roof (in house visual inspection)
Smoke detectors
Sump pumps
Water softeners

Quarterly
Air conditioners
Chiller—inspection by service company
Condensate drains
Cooling tower—complete inspection, grease motors
Dishwashers
Exhaust fans
Fire sprinklers—quarterly sprinkler inspection
Furnaces (in house inspection)
Grease traps
Refrigerators and freezers (commercial reach-ins or walk-in boxes)
Retention ponds
Smoke detector battery change

Semi-annually
Automatic air eliminators on heating/cooling systems
Backflow prevention device testing (RP type only) by qualified tester
Carbon monoxide detectors (change batteries)
Fire alarm (by certified professional)
Kitchen fire suppression system testing by certified or licensed contractor
Rain gutters
Resilient floors—strip, seal, and refinish
Smoke detector batteries (usually October and August)
Sweep parking lot
Water valves

Annually

Backflow prevention devices—testing by certified tester, drain to prevent freezing in winter
Boiler/water heater inspection (certified inspector)
Chiller—complete system inspection by service company
Electrical systems—visual building wide inspection
Elevator—3rd party inspection
Emergency generator—annual service by service company
Emergency lighting and exit signs
Fire extinguisher inspection by licensed contractor (annual maintenance)
Fire hydrant—flow test
Fire sprinklers—annual sprinkler inspection
Furnaces (complete service by qualified professional)
Ice machine—change water filter
Lawn irrigation systems—shut down system in the fall, start up in the spring
Lockers—change combinations and repair
Municipal fire inspections
Parking lot (seal coat)
Retention ponds
Roof (inspection by installing contractor)
Swimming pool seasonal startup and shutdown

Bi-annually or Less Often

Automatic air eliminators—replace every 5 years
Carbon monoxide detectors—replace every 2 years
Electrical bonding inspection of swimming pools every 5 years.
Electrical distribution systems—thermal imaging
Fire extinguishers, 5-year hydrostatic testing for dry chemical extinguishers. 6-year inspection and 12-year hydrostatic testing for CO_2 and wet chemical extinguishers
Fire hydrants—pipe flow test
Lockers—have lockers repainted
Re-stripe parking lots every 5 to 10 years.
Roof inspection by independent 3rd party every 5 years
Roof—asphalt or tar seal coat every 3 to 5 years (if required by roofing warranty)

Undetermined

Automated external defibrillators

Carpets—extract or shampoo carpets determined by the amount of foot traffic.

Septic tanks—depend on usage.

Termite inspections—depends on region of country.

Vehicles—varies depending on model.

Water filter elements. Based on volume of water used, concentration of contaminants in water, and size of filters.

LIST 2: PM TASKS REQUIRED BY BUILDING CODES OR REGULATORY AGENCIES

Many PM tasks are required by regulation or building code. Codes often vary by location. The following list includes the specific codes and regulations that apply in most jurisdictions in the US. Most often, these specific standards are not a written part of a local building code but are referenced by the locally adopted code. You will need to check with your local, county, or state agencies to find out if other codes or standards apply in your location.

In addition to differences in local jurisdictions, many industries will have industry specific regulations or even industry specific regulatory agencies that can alter this list. Health care facilities will often have additional requirements enforced by a state department of health while Schools will often have additional specific regulations created by a state departments of education.

The following list may need to be adjusted for your specific location and industry. For PM tasks, or frequency see the specific equipment description listed in this chapter alphabetically.

Backflow Prevention Device Testing—*EPA and National Standard Plumbing Code*

Boilers/Water heaters—*ASME Boiler and Pressure Vessel Code*

Carbon Monoxide Detectors—*NFPA 720 Standard for the Installation of Carbon Monoxide (CO) Warning Equipment in Dwelling Units*

Doors (smoke and fire doors) inspect and adjust—*NFPA 101 Life Safety Code*

Elevators—*ASME A17.1—Safety Code for Elevators and Escalators, ASME*

A17.2.1 (or ASME A17.2.2)—Inspectors Manual for Electric (or Hydraulic) Elevators, ASME A17.3—Safety Code for Existing Elevators

Emergency Generators—*NFPA 101 Life Safety Code, NFPA 110 Standard for Emergency and Standby Power Systems*

Emergency Lighting and Exit Signs—*NFPA 101 Life Safety Code*

Exhaust Fans—*ASHRAE 62 Ventilation for Acceptable Indoor Air Quality*

Fire Alarm System—*NFPA 72 The National Fire Alarm Code*

Fire Extinguisher Inspections—*NFPA 10 Standards for Portable fire Extinguishers*

Fire Hydrants—NFPA 25 *Standard for Inspection, Testing, and Maintenance of Water-Based Fire Protection Systems*

Fire Sprinkler Systems—NFPA 25 *Standard for Inspection, Testing, and Maintenance of Water-Based Fire Protection Systems*

Kitchen Exhaust Hoods and Fans—*NFPA 96 Standard for Ventilation Control and Fire Protection of Commercial Cooking Operations*

Kitchen Fire Suppression Systems—*NFPA 96 Standard for Ventilation Control and Fire Protection of Commercial Cooking Operations, NFPA 13 Standard for the Installation of Sprinkler Systems, NFPA 17 Standard for Dry Chemical Extinguishing Systems, NFPA 17A Standard for Wet Chemical Extinguishing Systems*

Playground Equipment—*CPSC Handbook for Public Playground Safety*

Smoke Detectors (battery operated)—*NFPA 72 The National Fire Alarm Code*

Code Writing and Regulatory Agency Contact Information

NFPA—National Fire Prevention Association
1 Batterymarch Park,
Quincy, Massachusetts
USA02169-7471 Tel: (617) 770-3000
www.nfpa.org

ASHRAE—American Society of Heating, Refrigeration, and Air conditioning Engineers
1791 Tullie Circle, N.E.
Atlanta, GA 30329
Tel: (800) 527-4723
www.ashrae.org

ASME—American Society of Mechanical Engineers
Three Park Avenue,

New York, NY 10016-5990
Tel: (800) 843-2763
www.asme.org

EPA—U.S. Environmental Protection Agency
Ariel Rios Building,
1200 Pennsylvania Avenue, N.W.
Washington, DC 20460
Tel: (202) 272-0167
www.epa.gov

CPSC—Consumer Product Safety Commission
4330 East West Highway,
Bethesda, MD 20814 Tel:
Tel: (301) 504-7923
www.cpsc.gov

LIST 3: EQUIPMENT SPECIFIC PROCEDURES, REQUIREMENTS, AND TECHNICAL DETAILS FOR PM

The following is an alphabetical listing of roughly 100 different types of equipment that should be considered for inclusion in your preventive maintenance program. Under the heading for each type of equipment are suggested PM procedures and PM frequency, information on applicable building codes or regulations, and technical information useful to the PM technicians that will be performing the actual preventive maintenance.

In this section, four symbols are used to provide additional information about each type of equipment listed. The four symbols and their meanings are:

INFORMATIONAL SYMBOLS

—The ambulance symbol indicates that maintenance of this type of equipment may effect the health and safety of building occupants.
—The earth symbol indicates that maintenance of this type of equipment may have an impact on the environment or may be governed by

environmental regulations.

🏛—The government building symbol indicates that maintenance of this type of equipment may be regulated by codes or government regulations.

☎—The telephone symbol indicates that maintenance of this type of equipment often requires calling an outside contractor instead of performing the work in-house.

AEDs

See Automated External Defibrillators

AIR COMPRESSORS 🌎

The most common uses for air compressors in facilities are as a shop tool in the maintenance department, to provide air for older pneumatic HVAC controls, or to provide air for dry pipe fire sprinkler systems. All of these types of Compressors require similar maintenance.

PM Tasks

There are three maintenance tasks that can significantly extend the useful life of an air compressor. These are maintaining proper lubrication, keeping standing water from accumulating in the pressure tank, and keeping air intake filters clean.

Compressor crank case oil will rarely, if ever, need to be changed. Unlike the oil in the crank cases of internal combustion engines, compressor oil is not exposed to soot, carbon, and other products of combustion and is not subject to the high operating temperatures found in engines. Compressor oil will therefore last for a very long time, most likely the life of the compressor. The oil level should be checked monthly.

Humidity in the air tends to condense as liquid water inside the pressure tank of air compressors. This water can contribute to rust which can lead to tank failure or rust particles clogging devices in the air stream. Water should be purged from compressors at least monthly.

Intake filters protect compressor bearing surfaces from wear caused by dirt particles. As with air conditioners, dirty air tends to bypass clogged filters and enter the compressor where it can cause abrasive damage and contaminate lubricating oil. Air filters should be inspected, and cleaned if necessary monthly.

Recommended PM Frequency

Monthly: Check oil level, purge water from tank, inspect and clean air filters.

Technical Notes

Automatic bleed devices are available from most industrial suppliers that are installed at the tank drain petcock to bleed off any water automatically. These simple devices open a drain valve for a few seconds every day.

Air compressors that are part of your dry fire sprinkler system should be serviced with your fire sprinkler system. See Fire Sprinkler Systems for more details.

AIR CONDITIONERS 🌎 ☎

Few kinds of equipment benefit from PM as much as air conditioners. Preventive maintenance will maintain their efficiency, extend the equipment's life, and have a dramatic effect on the bottom line. AC PM is important enough that you can find an entire section in this book on the technical aspects of maintaining air conditioners in Chapter 7. Be sure to read that section when setting up your PM system or when performing air conditioner PM.

PM Tasks

Quarterly: (during operating season) Change filters. Check for oil spots (indicating leaks). Check condensate pan for corrosion or leaks. Check condensate drain line and operate condensate pump. Grease blower and motor bearings. Verify operation of compressor. Check condition of drive belts.

Annually: (at start of cooling season) Clean condenser and evaporator coils. Verify operation by running unit. Check superheat to verify proper charge (requires EPA license). Check condition of drive belts and pulleys.

Recommended PM Frequency

Quarterly: minor maintenance (see section above)
Annually: major maintenance (see section above).

Technical Notes

See Chapter 7—HVAC Systems.

See Condensate Drains, Chillers, Cooling Towers

AIR ELIMINATOR

See Automatic Air Eliminator

AIR VENT

See Automatic Air Eliminator

ANSUL SYSTEMS

See Kitchen Fire Suppression Systems. Fire suppression systems from all manufacturers are sometimes incorrectly referred to as "Ansul" systems. Ansul is a subsidiary of Tyco Fire and Safety.

AUTOMATED EXTERNAL DEFIBRILLATORS

Automated defibrillators were almost unheard of a decade ago but are becoming commonplace in many facilities today. Since AEDs first started being accepted as crucial life saving equipment in facilities a few years ago, many lives have been saved with these devices.

Figure 11-1. Automated External Defibrillator

PM Tasks

Each manufacturer will have very specific maintenance requirements that must be followed. AEDs are largely self diagnostic and will alarm if the batteries are low or if there are other problems with the units. Inspections usually include making sure that the paddles, gloves, CPR mask, and other items are present, checking the units display or status indicator lights, and making sure that the unit is accessible and that there is appropriate signage in place throughout the building indicating the location of the AED.

Recommended PM Frequency

Follow the manufacturer's requirements

Technical Notes

These medical devices can deliver a controlled electric shock to help someone experiencing a heart attack. AEDs are designed to be used by first responders with little training and are able to diagnose cardiac arrhythmia (electrical problems with the heart) and provide a life saving jolt of electricity.

Many facilities now have AEDs located in areas where many people are likely to congregate or in areas such as on-site gyms where people experience physical exertion. AEDs are usually brightly colored and wall mounted in distinctively recognizable cabinets similar to fire extinguisher cabinets.

AUTOMATIC AIR ELIMINATOR

Many facilities have boilers or chillers to heat or cool water which is circulated throughout the building to unit ventilators, duct coils, or other heating or cooling appliances to condition building air. In buildings that use heated or chilled water for temperature control, there can be no air in the closed circulating loop. Air in the system will prevent circulating pumps from working efficiently and can block water flow. When air pockets prevent the flow of water in a heating or cooling, the system is said to be air bound. Air in heating systems also leads to corrosion of pipes, pumps, and other equipment.

Air can enter a heating or cooling system when boilers or chillers are opened for inspection or service, when circulating pumps are repaired, or when water with dissolved gasses enter the system as make up water.

Automatic air eliminators, also called automatic air vents are simple devices located at the points of highest elevation in the system

Figure 11-2. Automatic Air Eliminator

(where air will collect) that allow air to exit the system but prevent water from leaking.

PM Tasks

Automatic Air Eliminators often fail due to corrosion or age. A failed air eliminator may fail in the closed position and not allow air to escape or they can fail in the open position resulting in a fairly significant water leak.

Air vents should be tested at the start of the heating season for heating systems and should be tested at the start of the cooling system for chilled water systems. When both systems are used, air vents should be tested twice a year.

Air vents are very inexpensive, costing a few dollars each. Many heating and cooling specialists recommend replacing automatic air eliminators every 5 years.

Recommended PM Frequency

Annually: test by depressing valve stem to make sure air vent will open at start of heating season for hot water heating systems.

Annually: test by depressing valve stem to make sure air vent will open at start of cooling season for chilled water heating systems.

Every 5 years: replace all air vents in the facility.

Technical Notes

Most automatic air eliminators have an internal float that closes a small needle valve when the float is lifted by water in the heating system and opens the valve when the float drops whenever air is present.

Whenever replacing air eliminators, it is good practice to install a ball valve between the air eliminator and the heating or cooling loop. Adding this very inexpensive part can make a very big difference in preventing a flood if an air vent should fail in the open position. Having such a valve will make replacing a faulty air vent much faster and easier in the future.

Larger heating/cooling systems may include air separators in addition to or as part of an air vent assembly. Water flows through a large chamber in an air separator where the water velocity is slowed. Various structures are used inside the air separator chambers to cause micro bubbles of air to combine into bubbles that are large enough to float to the top of the chamber where they are released through an automatic air eliminator.

AUTOMATIC AIR VENT

See Automatic Air Eliminator

AUTOMATIC TRANSFER SWITCHES (ATS)

See Emergency Generators

AUTOMOBILES

See Vehicles

BACKFLOW PREVENTION DEVICES

Backflow prevention devices are designed to protect drinking water from sources of contamination. Backflow devices are often required between municipal water service and individual buildings or between buildings and sources of possible contamination such as lawn irrigation systems, boilers, fire sprinkler systems, chemical treatment systems, and others. Contamination of water can occur if potable (drinking) water pressure were to drop allowing contaminated water to flow back into the potable water supply.

PM Tasks

The National Standard Plumbing Code and the EPA require testing of backflow protection devices annually. Some jurisdictions require RP type devices to be tested semi-annually. Depending on your local jurisdiction, testing must be done by either a licensed plumbing contractor or by a certified backflow device tester.

Recommended PM Frequency

Annually: Test proper operation of AVP and PVB devices.

Annually: Drain outdoor backflow preventers to prevent freezing damage.

Semi-annually: Test proper operation of RP devices.

Technical Notes

There are four basic types of backflow prevention devices in use.

1. *Double Check Valve*

   The simplest type of back flow prevention device is the double check valve. As the name implies, this device is made of two check valves in series. Using two check valves instead of one provides redundancy and an added level of protection. Many jurisdictions do not allow double check valve back flow preventers to be used to protect the public water supply. Where they are allowed, they are not suitable for high hazard applications.

2. *Atmospheric Vacuum Breaker*

   The simplest type of backflow device approved for most applications is the atmospheric vacuum breaker (AVB). An AVB includes a vent at the top to allow air to enter the system if a siphon should develop and a floating poppet which seals off this vent when water pressure is present. If water pressure should drop, the poppet will fall opening the air vent preventing contaminated water from siphoning into the potable water side of the a device. For AVPs to work properly, they must be installed at least 6″ above the highest water level downstream from the device. AVPs are only suited to applications where the hazard of contamination is low. The Hose bibb vacuum breaker (HBVB) is a simple type of AVP that is installed onto a hose bib or faucet to prevent backflow from garden hoses into a building's water system.

3. *Pressure Vacuum Breaker*
   The pressure vacuum breaker (PVB) type of backflow device is similar to the AVP except that the poppet is spring loaded. The poppet spring ensures that the poppet will open before water pressure drops to atmospheric pressure. Approved PVBs will also include a test cock (a small valve with a port for attaching a set of testing gauges) on both the inlet and outlet side of the device and two isolation valves. PVBs can be installed in locations where there is a high hazard of contamination and they may have valves installed downstream of the device.

4. *Reduced Pressure Backflow Prevention Assembly*
   A reduced pressure (RP) backflow prevention assemblies are also known as reduced pressure zone (RPZ) or reduced pressure principal (RPP) backflow devices. These devices include two check valves with a pressure relief valve between the two check valves. The check valves provide redundant protection against flow in the wrong direction while the pressure relief valve provides protection if the check valves were to fail. RVPs also include a test cock before

Figure 11-3. Faucet with Atmospheric Vacuum Breaker

Figure 11-4. Pressure Vacuum Breaker (PVB)

the first check valve, between the two check valves, and after the second check valve to allow the device to be tested. RPVs are used where a high hazard of contamination exists.

BACKUP GENERATORS

See Emergency Generators

BOILERS AND WATER HEATERS

Regulations vary depending on the size of the boiler (in Btu or Hp), whether the boiler produces hot water or steam, and the boiler's operating pressure. In most jurisdictions there are several classes of boiler licenses (called seals) required to operate certain types of boilers.

PM Tasks

Operating temperatures and pressures need to be monitored regularly, often every two hours for larger boilers. The operation of safety

Figure 11-5. Reduced Pressure Backflow Preventer

devices such as low water cut outs and safety valves should be tested for proper operation daily or on every shift.

Combustion burners (oil or gas) should be cleaned and adjusted every year, at the start of the heating season by someone qualified to do this work. Oil nozzles and oil filters should be replaced, oil pump pressure should be tested, and the combustion chamber should be cleaned and inspected. Gas appliances need to be cleaned and burners adjusted for the most efficient operation. Barometric dampers should be adjusted. Water tube and fire tube boilers should be opened for inspection annually.

Inspections by certified service companies are often required as well as regular inspections by regulatory agencies for safety issues and environmental impact. Environmental permitting may be required to operate boilers.

Recommended PM Frequency

Note: Frequency of PM tasks is different for different size boilers. If you are performing PM on a boiler with specific requirements, you should have the appropriate operating license for that boiler and will be familiar with the requirements. The following requirements are generic for smaller boilers and hot water heaters.

Monthly: If a boiler was to malfunction and overheat, there are two devices that will prevent a boiler explosion, the safety relief valve and low water cut off. Test the safety relief valve by manually lifting

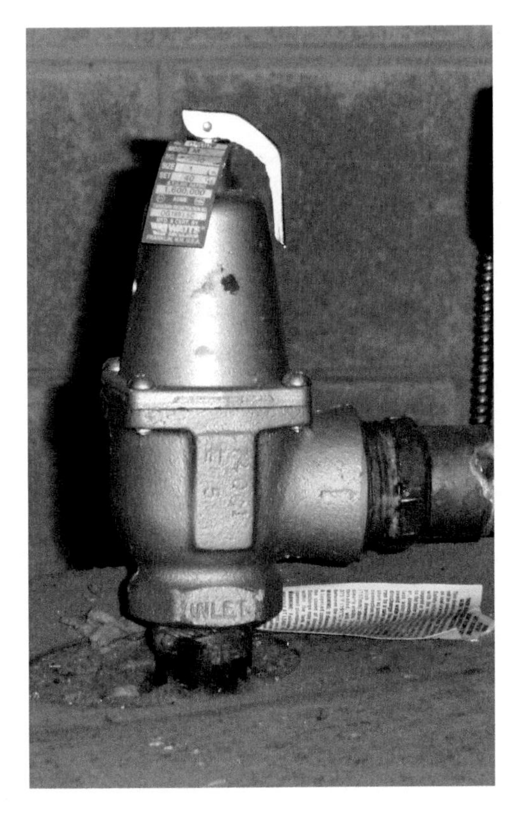

Figure 11-6. Safety relief valve. Test by lifting handle until water runs clear.

the handle until water runs clear. Test low water cut off by opening LWCO drain or following manufacturer's instructions. Test these devices at least monthly. Many jurisdictions require daily testing for larger boilers.

Annually: Boilers should have a safety inspection by a certified inspector or by your regulatory agency if required. Combustion boilers should be inspected by a qualified company for operation prior to start of heating season.

Technical Notes

One often overlooked item on small tank type hot water heaters is the anode rod. To prevent the hot water heater tank from corroding, this sacrificial rod is installed at the factory. The anode rod is designed to corrode to protect the tank. This rod should be inspected with your annual inspection and replaced if significantly corroded.

See also Furnaces.

CARBON MONOXIDE (CO) DETECTORS

CO is a colorless, odorless gas produced by the incomplete combustion of fossil fuels such as oil, natural gas or propane. CO is poisonous and at low exposure levels can produce symptoms similar to the flu. At high concentrations, death can result. Approximately 250 people die in the US from CO poisoning each year.

Many jurisdictions require CO detectors to be installed in sleeping units (hotel rooms, patient rooms, apartments) whenever combustion appliances are present.

PM Tasks

CO detectors should be tested monthly, usually by pressing the "test" button on the detector. Detector batteries should be changed twice a year, traditionally in October and April. Most manufacturers agree that CO detectors sensors have a design life of only 2 years. Consequently, CO detectors will need to be replaced every 2 years.

Because CO detectors are similar to Smoke Detectors in their maintenance requirements and location, Smoke detectors and CO detectors are normally included on the PM calendar together.

Recommended PM Frequency

Monthly: test detectors

Semi-annually: change detector batteries

Every 2 years: replace detectors unless manufacturer recommends a different replacement interval.

Technical Notes

Installation and maintenance of CO detectors is regulated by NFPA 720: "Standard for the Installation of Carbon Monoxide (CO) Detection and Warning Equipment.

CO detectors have a replaceable "sensor kits" that can be replaced instead of having to purchase new CO detectors every two years. When pressing the "test" button on a CO detector, you are probably only testing the detectors electrical circuits, not the condition of the sensor. If a detector is older than 2 years old, it may test fine and still have a bad sensor.

CARPET CARE

Carpet care often falls under the scope of the janitorial or house keeping department. However, there are many buildings where the lines between maintenance and custodial work are blurred. In some instances, the maintenance and housekeeping department are one and the same. In other instances, housekeeping may handle daily vacuuming of carpets with maintenance handling "heavy housekeeping" tasks such as carpet extraction when it is needed.

PM Tasks

The first and often overlooked step in carpet care is to keep the dirt from reaching the carpet in the first place. Keeping dirt out of your building will help to extend both the life of your carpets and the time between cleanings. Walk-off mats at entrance ways can have an enormous impact on the amount of dirt entering your building. A good system of walk off mats will knock off dirt, absorb water, and finally wipe shoes clean of any oils that that will yellow and attract more dirt to your carpet. A walk-off mat will need to be long enough for several steps to be made on the mat to be effective.

Any dirt in a carpet will act as an abrasive and cause carpet fibers to wear under foot traffic. Dirt particles are often microscopically jagged and will damage carpet fibers. This damage is often seen as carpet that appears dull or dingy. Once this damage is done, there is no amount of cleaning that can restore the carpet's appearance. Frequent deep vacuuming not only makes carpet look better, it actually extends carpet life. A well maintained vacuum with a brush beater will do the best job at removing dirt from within the carpet. Frequent vacuuming can extend carpet life by 50% in high traffic areas.

Spills and stains on carpet should be cleaned up as soon as possible. Stain spotter kits are available from most janitorial suppliers that will have 6 or 8 different spotting chemicals in spray bottles for different types of stains. The longer a stain is allowed to remain, the stronger the stain will be bound with the carpet fibers.

To remove abrasive soil deep within the carpet fibers, some sort of deep cleaning should be performed frequently. This may be steam cleaning, hot water extraction, or some other method that gets down to the bottom of the fibers and lifts the hidden dirt away.

Carpets are usually deep cleaned when they start to look bad. By the time carpets start to look like they need deep cleaning, damage has already been done. Carpets should be cleaned on a regular schedule before carpets appear soiled.

Wet carpets will attract dirt so it is important to dry carpets immediately following deep cleaning and to keep all traffic off of carpets until they are dry. Carpet drying can be accelerated by repeated vacuuming using the extractor, using drying fans, and by using a drying bonnet on a floor scrubbing machine, In areas where traffic cannot be avoided, try extracting only half the carpet each day and direct traffic around wet carpet.

Recommended PM Frequency

The amount of traffic on your carpets dictates how often they will need to be vacuumed and deep cleaned. Remember that both of these tasks should be performed more frequently than the carpet's appearance dictates.

Technical Notes

See also Resilient Flooring Care

CARS

See Vehicles

CAULKING

Caulking is used in many different ways in buildings and will show up in several different ways on your PM calendar. Caulk is used anywhere that a water or air tight seal is needed between two building components. Exterior building expansion joints will need to have the caulk maintained regularly, as does caulking around windows and doors. In hotels, a lot of time can is often spent by maintenance departments to keep the caulk around tubs and showers mildew free. Caulk joints show up around pipe or electric penetrations in exterior walls and at transitions between different roofing components. Caulk is the do-all product wherever there is a crack or gap.

PM Tasks

In most cases, caulk will last a very long time. Excessive movement caused by thermal expansion, wind, or building settling can cause caulk joints to fail. Humid conditions cause caulk to mildew, which is more of a cosmetic problem than a structural one.

PM involves inspection of critical caulk joints such as around penetrations or at building expansion joints. In the real world, inspecting all of the caulk joints around windows and doors would prove to be more time consuming than the benefits would justify unless you have an area with a history of caulk failures.

Having spent several years of my career in the hotel industry, mildewed caulk in bathrooms is one of my pet peeves. Cutting out and replacing mildewed caulk is a common task for the maintenance departments of most hotels. The time spent on mildewed caulk can be practically eliminated if the housekeeping department is given the proper tools, cleaning products, and held accountable for keeping the tub caulking mildew free every day. And reducing the work of the maintenance department is one of the biggest reasons to have a PM program.

Recommended PM Frequency

Annually: inspect building exterior expansion joints, penetrations, and other caulked components.

With Room PM: inspect any caulking in tubs, around AC units, windows, and wallpaper seams.

Technical Notes

There are several types of caulk available, all with different characteristics. When choosing a caulk for your particular job, you will need to consider the following 4 items:

1. Will you be using the caulk indoors or out? Some types will work better in one location than the other.
2. What will be the temperature? Some caulks can handle extreme temperatures while other can't.
3. Moisture Levels. Some caulks perform better in moist environments than others and may even have additives to help prevent mildew and mold from growing.
4. Application method. Some caulks are easier to apply than others and the skill level of the applicator should be considered.

Silicone Caulk

Silicone caulk has become one of the most common types of caulk over the past few years. It is available in many colors and has a rubbery, flexible texture when cured. Silicone caulk tends to stick to surfaces well and is especially useful on non-porous surfaces, such as metal and plastic. Since silicone remains flexible, it rarely cracks due to flexing.

Silicone is one of the most durable caulks lasting 30 years or more and stays flexible in all temperatures. Silicone caulk will not hold paint as paint tends to just slide off of the slippery surface. Silicone requires solvents for clean up and a mistake or smudge can be impossible to remove. Silicone caulk on hands often has to wear off after a couple of days.

Latex or Acrylic Caulk

Latex and acrylic caulks are very similar and are often blended as a latex acrylic formulation. Latex caulk is easy to apply, easy to clean up, and can last 20 years. Latex tends to remain flexible, but not quite as flexible as silicone. Latex caulk can be painted and is the most common choice for sealing interior cracks prior to painting.

Latex caulk does not hold up to temperature extremes and becomes very hard and brittle when cold. Latex is a poor choice for exterior caulking in cold climates.

Acrylic Latex Silicone Blend Caulk

This type of caulk is a blend of acrylic, latex, and silicone caulks and has characteristics of all three types. Acrylic latex silicone caulk is easier to apply than silicone but has similar durability and flexibility. Some formulations can be painted while other's can't. This type of caulk is good for use indoors or outdoors in all weather conditions.

Kitchen and Bath Caulk

This type of caulk is usually one of the previous types with chemicals added to prevent mildew growth. Characteristics will be similar to the types mentioned above.

Butyl Rubber Caulk

Butyl rubber caulk is probably the most adhesive of any of the caulks. This characteristic along with the fact that Butyl rubber is extremely elastic and flexible makes it an excellent choice for use in large gaps and cracks. Strong and durable, butyl rubber caulk can last up to ten years outdoors.

Because butyl rubber caulk is so sticky, cleanup is nearly impossible. Any drips or application mistakes are best chipped away after the caulk has had several days to cure. Caulk left on caulking guns or other tools will be a permanent addition.

Oil-based Asphalt Caulk

Although probably not an actual caulk, asphalt or tar-based sealants are available in standard caulking tubes for use in caulking guns and in buckets to be applied with putty knives or trowels. Asphalt sealant is often used to patch or seal around skylights, pitch pockets, chimney flashing, an other roof components. Asphalt sealant tends to take a very long time to dry and remains sticky for months. Asphalt and tar sealants should not be used where anyone might come into contact with the material. Asphalt-based sealants should only be used with asphalt-based materials and tar-based sealants should only be used with tar-based materials. For information on mixing tar and asphalt see Chapter 6 on commercial roofing.

CHILLERS

Chillers are large cooling machines that chill water. This chilled water is then pumped throughout the building to cooling coils to provide cooling in all areas. An air conditioner with a capacity of 30 tons of cooling would be considered a large air conditioner while chillers are often sized in the hundreds of tons of cooling capacity. A building that would require dozens of large air conditioner units can easily be served with a single chiller machine.

PM Tasks

Probably the largest disadvantage chillers have over conventional air conditioners is that an equipment failure can shut down the cooling for an entire building. Because of this, maintaining chillers should be a very high priority on your to do list. The cost of chiller repairs can also be staggering. Proper maintenance can prevent repairs that can easily cost tens of thousands of dollars. PM performed on reciprocating chillers has an average return on investment (ROI) of 400% and PM on centrifugal type chillers has a return of over 1,000%.

As a rule, I believe in handling as much PM work in-house as possible. Chiller plants are one of the exceptions to that rule. Chillers are expensive, costing upwards of $1,000 per ton of cooling capacity. You want factory trained service people who are familiar with your specific equipment, and who know what problems are likely to occur. The cost of a PM service call will more than be saved in downtime, repair costs, and extended equipment life.

Scale buildup on evaporator tubes reduce the efficiency of heat transfer. Ask your service company about chemical treatment to prevent scale buildup and how often they recommend having the tubes cleaned.

There are many different brands and types of chiller plants. As with all equipment, the manufacturer's PM instructions should be reviewed when setting up your PM program. The following PM tasks are generic.

Recommended PM Frequency

Daily: Observe, and listen to system operating. Check computerized operating log and history for any unusual conditions. Most chillers today are largely self-diagnostic and will notify you of any problems.

Weekly: Inspect sight glasses for oil level, operate lag pumps or switch lead/lag pump position to be sure all pumps will work if needed. Inspect piping for leaks.

Quarterly: Have your service company perform an inspection to include electrical connections, safety controls, superheat/subcooling calibration, and to check for refrigerant leaks.

Annually: Have your service company shut down system and take an oil sample for lab analysis, check the entire refrigerant system for leaks, test the condition of motor windings and insulation using a megohm-meter, and calibrate all safety and operating controls. Your service company may recommend vibration analysis to test the condition of compressor bearings.

Technical Notes

See also Chapter 7—HVAC Systems

CIRCUIT BREAKERS

See Electrical Systems

CIRCULATOR PUMPS

Circulator pumps move water. They are found in heating and cooling systems, used to move domestic water long distances, used to increase pressure or to move water to the tops of buildings.

PM Tasks

Check motor ventilation openings and clear away any dirt blocking free movement of air. Lubricate bearings with the proper grease or oil. Check condition of rubber resilient motor mounts if used. Check couplers for wear, wobble, or shaft misalignment. Check shaft seal for leakage.

Figure 11-7. Bank of circulator pumps

Recommended PM Frequency
Monthly

Technical Notes

Circulator pumps typically include three components: The pump, a driving motor, and a coupler/bearing assembly connecting the two together. Some designs connect the motor shaft directly to the pump impeller. All three components are included when performing PM to circulator pumps.

COOLING TOWERS

Cooling towers are the "other half" of chiller plants. Chillers produce chilled water by extracting heat from the chilled water loop running throughout the building. This heat is often transferred to a second water loop, the cooling tower loop where the heat is taken outside and released to the environment. The cooling tower loop is sometimes referred to as the condensate loop.

PM Tasks

Since a failure of the cooling tower can shut down the cooling for an entire building or several buildings, PM of cooling towers should be near the top of your PM priority list.

Cooling towers are fairly simple pieces of equipment and a quick daily inspection should turn up most small problems while they are still small.

Recommended PM Frequency

Daily: Visual inspection of the cooling tower in operation, check the water level in the water reservoir, make sure fans are operating. You should be noting the condensate temperature each day when doing your daily chiller inspection. (see the section on chillers.)

Weekly: Operate each fan independently and listen for any sounds or problems. Operate circulating pump. Check water level and manually operate make up water valve. Check for leaks.

Quarterly: Inspect and change any clogged spray nozzles, check fan motor belts, grease fan motors and circulating pumps.

Technical Notes

There are many different cooling tower designs. In a typical cooling tower, hot water from the chiller machinery is pumped outside to the cooling tower where it is cascaded down through a series of baffles. Fans draw ambient air across the hot water as it cascades down through the tower. As some of the water evaporates, it absorbs heat from the remaining water cooling it down for its return trip to the chiller plant.

There are several different designs of cooling towers. Many towers cascade and evaporate the water from the cooling tower loop down through the tower. Others keep the condensate water in a closed exchanger coil and cascade water down around the coil to cool the condensate water but the two water sources are kept separate. Still other cooling towers pump the condensate water through a direct exchange coil with only ambient air drawn through the coil to cool the condensate water.

CONDENSATE DRAINS

Condensate drains can refer to the piping or tubing that carries condensate water away from an air conditioner. Condensate drains can also refer to drains and drain traps in the building that collect this condensate and move it out of the building along with other waste water and sewage.

PM Tasks

For condensate drain lines from AC units, use compressed air to maintain clear lines. Add algae control tablets to condensate pans to prevent the growth of algae which will block condensate drains.

Recommended PM Frequency

Start of cooling season: Blow out condensate lines.

Quarterly: Check air conditioner drain pans and condensate lines during the cooling season. Add algae control tablets to condensate pan.

Monthly: If your building has dedicated plumbing drains to handle condensate water, pour water into these traps during heating season to maintain a water plug against sewer gasses.

Technical Notes

There are tools made specifically for blowing out condensate lines. These take a tiny compressed air cartridge similar to those used in air rifles. Any portable air tank, portable compressor, or even nitrogen tank (with a regulator) will work. Condensate drain lines can be blown from the AC out or can be blown toward the AC from the discharge end.

Condensate lines on many AC units have a plumbing trap close to the air handler as part of the condensate line. When the evaporator coil is located before the fan in an air handler, the fan draws air from the evaporator coil causing the condensate line to be under negative pressure. This negative pressure will pull air through the condensate line preventing water from flowing away. Eventually the backed up water will cause the condensate pan to overflow. A 4″ high drain trap in the condensate line creates a water plug that prevents the fan from pulling air through the condensate drain line. Condensate traps are also places where clogs are common.

There must be an air gap between a condensate line from an AC unit and the building condensate trap it empties into. If a condensate drain were to be connected directly to the building's drain system; sewer gasses could be drawn into the AC system and would be distributed throughout the building through the ductwork.

See Floor Drains in this chapter for tips on using baby oil or barrier traps to make condensate drain maintenance easier.

See Air Conditioners.

DEFIBRILLATORS

See Automated External Defibrillators.

DETENTION PONDS

See Retention Ponds

DISHWASHERS

Commercial food service kitchens rely on large, automated dishwashers for ware washing.

PM Tasks

Dishwasher temperatures and chemical concentration readings need to be measured and recorded daily. Your local health department will probably want to see these records every time they inspect your kitchen. Daily temperature and chemical readings are probably already being done by the kitchen staff and but should be verified during PM inspections.

During the PM inspection, check for leaks and water under the dishwasher, check door seals and the condition of spray shields on the inlet and outlet side of the tray conveyor. Check spray, drain and chemical pump operation.

Check booster heater and dishwasher heating element current draw. It's not unusual to find that one or two heating elements have failed and that the remaining elements are running for longer intervals to catch up.

The dishwasher drain valve must seat fully or high temperatures will be hard to reach as cooler makeup water will be constantly added to the machine.

Recommended PM Frequency

Quarterly: Inspect and operate all components and systems.

Technical Notes

Most commercial dishwashers are one of two different types. The first is the chemical sanitizing dishwasher and the second is hot water

sanitizing dishwasher. Chemical sanitizing dishwashers use sanitizing chemicals to sterilize dishes. Chemical sanitizing machines require a rinse water temperature of at least 120°F and often a temperature of 140°F and are known as low temperature dishwashers.
Hot water sanitizing dishwashers rely on the temperature of the water to sterilize dishes. Hot water machines require a temperature of at least 180°F and are called high temperature dishwashers.

Water is usually supplied to the dishwasher at 140°F for the wash cycle and built in heating elements in the machine help to maintain this temperature. Booster heaters are used increase the temperature of the water to 180°F for the rinse cycle.

DOORS

Entrance doors can be subject to an enormous amount of wear and tear. Regular minor adjustments to door closers, strikes, and hinges will keep doors operating for a long time.

Smoke and fire doors are used to contain smoke and fire from spreading through a building. *NFPA 101—The Life Safety Code*, and local construction codes require that these doors must close and latch on their own and must maintain a tight seal between building compartments.

PM Tasks

Let doors swing closed on their own and verify that doors latch properly. Check hinges, panic hardware, and other hardware for tightness. Adjust closer speed so that doors latch consistently but do not close fast enough to injure a child. If magnetic hold-opens are used, verify that they release automatically during a fire alarm when performing your monthly fire alarm tests. Check operation of door synchronizers on double doors.

Recommended PM Frequency

Monthly: inspect and adjust entrance doors, smoke and fire doors, and any doors subject to frequent use.

Technical Notes

Most manufacturers' door closers are adjusted with two or three screws located on the body of the door closer. The screw marked 'B' adjusts the speed of the backswing or door opening speed. The screw marked "S" adjusts the swing speed of the door as it is closing. If an "L" screw is available, this adjusts the latch speed which is the last few inches of the door's swing just before the door latches.

Electromagnetic door release devices hold smoke and fire doors open until the fire alarm system activates. During a fire alarm, power is dropped to the electromagnets releasing the doors and causing them to close and latch automatically. Doors in corridors or to stairwells are often held open using these devices. Be sure to test the operation of these magnetic releases when performing the monthly test of you building's fire alarm system.

DRAINS

See Floor Drains, Condensate Drains

DRYERS (COMMERCIAL LAUNDRY)

See Laundry

ELECTRICAL SYSTEMS

A facilities electrical distribution system include everything between the power companies (POCO) service to your building and the branch circuits that feed individual receptacles, appliances, and specific pieces of equipment. Distribution equipment includes transformers, main distribution panels (MDPs), and all of the electrical subpanels that provide electricity to the branch circuits in a facility.

The most common problem with electrical distribution equipment is overheating caused by loose or corroded electrical connections. Circuit breakers and fuses are designed to open a circuit when excessive current travels through the breaker. They do this by reacting to high temperatures within the circuit breaker or fuse due to the excessive current. Loose or corroded electrical connections can overheat causing circuit breakers to trip and fuses to fail.

High temperatures can cause breakers to fail mechanically, can cause wire insulation to burn away and can cause plastic insulating components to fail.

PM Tasks

It's a good idea to perform a visual electrical inspection at least annually to look for electrical problems. The inspections should look for obvious electrical problems such as extension cords used as a replacement for permanent wiring, open junction boxes with exposed wiring, broken light fixtures, and missing ground fault circuit interrupters (GFCI) where required. A small receptacle tester can be used to check receptacles for reversed wiring, broken ground paths, or other electrical problems. GFCIs should be tested by pressing the "test" button on the face of the receptacle to force the receptacle to trip.

Thermal imaging is the fastest and most complete method of inspecting electrical equipment. Thermal cameras can see hot spots in electrical equipment that can indicate loose or corroded connections long before those hot spots cause damage. Thermal imaging requires removing the panel covers from electrical panels and viewing the panels while energized and equipment is operating. Since thermal imaging cameras cost several thousand dollars, most facilities hire electrical contractors or inspection companies to perform thermal imaging.

Recommended PM Frequency

Annually: Perform a walk through with a visual inspection of all electrical components in facility. Test receptacles with proper tester. Test GFCIs by pressing "test" button. Make repairs to any problems found.

Every 2 or every 3 years: Have entire electrical distribution system inspected with thermal imaging.

Technical Notes

Although not a "PM" activity, labeling individual branch circuits will have a big impact on continually improving maintenance activities. Figuring out which circuit breaker serves which receptacle, exhaust fan, or light fixture can be time consuming and usually requires two people who can communicate by radio. Knowing this information can save hours of searching in the future. If your facility has a set of as-built blueprints, the individual branch circuits will be listed there. Some changes have probably been made over the years but these prints will give you a head start on finding the circuits.

Circuit breakers are designed and tested to be extremely dependable and require very little preventive maintenance. Some industrial users test circuit breakers annually if they serve critical equipment. Testing procedures are extensive and are written by the National Electrical Manufacturer's Associations (NEMA). In critical areas of hospitals, data centers, emergency communication centers, or a few other types of facilities, annual circuit breaker testing may be required. The rest of us can consider molded case circuit breakers to be maintenance free.

Space heaters and extension cords are two common electrical violation cited during fire inspections. Space heaters are not allowed and extension cords are only allowed for temporary use. Most jurisdictions will allow a power strip with a built-in circuit breaker to be used instead of extension cords.

Probably the most common problems found during electrical system inspections are housekeeping issues including: storage in front of electrical panels, trash and debris in electrical rooms and on electrical equipment, and inadequate working space.

ELEVATORS

Elevators and escalators accidents kill roughly 30 and seriously injure roughly 17,000 people each year in the US. Most of these injuries involve car leveling problems where the car floor and landing floor are not aligned properly causing a tripping hazard.

ASME A17 allows the local Authority Having Jurisdiction (AHJ) to determine how often elevators should be inspected. Typically elevators require annual inspections by an independent third party. This means by someone who is not employed by the elevator owner nor by the elevator service company contracted to maintain the elevator. In most locations, only qualified elevator inspectors (QEIs) who have been certified by a QEI certifying agency can perform elevator inspections.

Although annual 3rd party inspections are usually required, monthly inspections should also be performed. Your elevator service company can provide a comprehensive inspection or you can perform a visual inspection in-house.

PM TASKS

The monthly in-house inspection should include the following:

- Make sure that the doors close with no more than a 3/8" gap at all landings.
- Verifying that the car levels evenly with each floor.
- Make sure the electric eye or safety edge door sensors operate correctly and that the doors will not close if someone is standing in the doorway.
- Check that all indicator lights light and all buttons operate.

- Make sure the car telephone reaches someone that can send help during all hours that the building is occupied. (This might be a reception desk or the local police department).
- Inspect the machine room for storage, clutter, and other housekeeping issues.
- Inspect the elevator pit of hydraulic elevators for signs of oil leaks
- Verify that the emergency lighting operates if there is a power outage.

Recommended PM Frequency

Monthly: In-house visual inspection

Annually: Inspection by certified elevator inspector

Technical Notes

There are two types of elevators that are used in nearly all buildings. The first type is the cable elevator where the elevator car is suspended by 5 or more steel cables and is raised and lowered by a mechanical winch at the top of the elevator shaft. Each of the 5 cables is capable of supporting the elevator's weight if the other 4 cables were to fail.

The second type of elevator is the hydraulic elevator. A hydraulic elevator is supported by a large hydraulic piston located below the building. These pistons are installed as deep in the ground as the building is tall.

EMERGENCY GENERATORS

Emergency generators provide electricity in the event of a disruption of service by the electrical utility. Emergency generators are rarely sized to provide for a building's entire electrical needs and will probably only provide enough electricity for critical equipment. This would include a percentage of lighting to allow safe exit from the building, the fire alarm system, refrigerators and freezers, the telephone system, and possibly a few convenience outlets. In healthcare facilities, electrical outlets for medical equipment, nurse call systems, and other life safety systems would be included in this list.

PM Tasks

NFPA 110 *Standard for Emergency and Standby Power Systems* requires quarterly testing of emergency generators. One of the four annual tests must be a test under load. However, most organizations choose to

Figure 11-8. Emergency generator

test their generators weekly or monthly to make sure the generator will run when needed. A full load test, meaning the building is switched over to generator power, is often performed monthly. A full load test usually does not effect any electrical circuits that are not part of the emergency electrical distribution system. Most building occupants will notice only a brief (less than a second) flicker of the lights and everything else will operate normally with no interruptions.

Most generators perform a weekly test automatically via a programmable transfer switch. The weekly test may or may not be a full load test.

A service company should be contracted to perform annual service on your generator. Annual service includes changing the oil, changing filters, topping off or replacing engine coolant, testing batteries, and other engine service items. The generator is run and tested for proper operation.

Recommended PM Frequency

Weekly: Generator Running Test

Monthly: Generator Full Load Test (can be eliminated if weekly test is under load)

Annually: Annual service by a generator service company

Technical Notes

Emergency generators are powered by internal combustion engines running diesel fuel, natural gas, or occasionally gasoline. These engines have the same maintenance requirements as other internal combustion engines. Oil changes need to be done annually or after a specific number of hours of operation, Batteries and their chargers need to be maintained, air and fuel filters need to be changed, belts need to be checked, The coolant level needs to be checked, as well as other items. Typically, a service company is contracted to perform a complete generator service annually which will include all of these items. If your generator's operation is critical, it is recommended that an oil sample is sent out for analysis to find any hidden problems.

Emergency generators provide their power to the building through a transfer switch which transfers the building from the electrical utility's power lines and onto the emergency generator. Transfer switches are required to prevent the electricity from the generator from entering the power utility's lines where it could injure or kill someone working on the electric lines. Most of these switches are automatic (Automatic Transfer Switches ATS) and will automatically start the generator and transfer power to the building within 10 seconds of a power interruption.

EMERGENCY LIGHTING AND EXIT SIGNS

Exit and emergency lights are necessary to allow safe evacuation during a fire or power outage. NFPA 101 is the accepted code in most jurisdictions.

PM Tasks

Battery operated emergency lighting and battery operated exit signs should be tested by pressing the "test" buttons on the equipment. Generator operated fixtures need to be tested to verify that they will come on under generator power.

Recommended PM Frequency

Monthly: verify that emergency lights will operate on battery power for 30 seconds.

Annually: test battery operated fixtures for 90 minutes.

Technical Notes

For fixtures that operate on an emergency generator, schedule inspections on the same date as your generator test to avoid doing double work.

According to *NFPA 101—Life Safety Code*, emergency lighting must be tested for 30 seconds every 30 days and must be tested for a 90 minute duration once a year.

EXHAUST FANS

Exhaust fans remove unpleasant odors from rest rooms, remove smoke and heat from commercial kitchens, and remove stale air from offices, classrooms, and other occupied spaces. In some areas, exhaust fans are also used to ventilate the ground below a building to prevent any accumulation of radon gas.

PM Tasks

Remove the cover from all exhaust fans and inspect any pulley's and drive belts for wear. Grease or oil bearings. Check fan blade for excessive wobble. Make sure motor runs. Repair any problems that are found.

Recommended PM Frequency

Quarterly

Technical Notes

The typical rate of air exchange designed into a building is 15 to 20 Cubic Feet per Minute (CFM) of fresh air for each occupant. Fresh air can only be brought into a building if old air is exhausted.

See also "Kitchen Exhaust Hoods and Fans."

EXIT SIGNS

See Emergency Lighting and Exit Signs

FANS

See Exhaust Fans

FIRE ALARM SYSTEMS

Fire alarm systems are one of the most important systems found in a building and must be maintained according to specific regulations.

NFPA 72—The National Fire Alarm Code is the code used in most jurisdictions.

PM Tasks

In house testing of the system should be done monthly by pulling a pull station or activating a smoke detector. A different pull station or detector should be used each month. Testing of the entire system will probably need to be performed by a licensed or certified service company semiannually or annually depending on local codes.

Recommended PM Frequency

Monthly: Test system operation by pulling a pull station or triggering a smoke detector. Verify that the alarm sounded and that your monitoring company received the alarm signal. Verify that horns sounded, strobes flashed, and electromagnetic door released closed smoke and fire doors.

Monthly: Schedule evacuation drills to correspond with your alarm tests. The required frequency of drills will vary with your location and type of building.

Semi-annually: Have a certified fire alarm company test your system. Typically 50% of smoke detectors are tested every 6 months so that 100% of your system is tested over the course of each year. NFPA 72 requires annual testing of most fire alarm systems.

Technical Notes

Before testing the fire alarm system, your monitoring company should be called and instructed to take your system "off line" or to place your system on "test" during the alarm. Your 911 dispatch office and local fire department should also be notified prior to activating the alarm.

When testing your system, choose a different pull station or smoke detector each month to verify that all zones are operating. Fire inspectors will ask to see documentation of fire alarm tests and inspections so keep excellent records.

FIRE EXTINGUISHERS

In most jurisdictions, portable fire extinguishers fall under the jurisdiction of *NFPA 10—Standard for Portable Fire Extinguishers*. The code requires several different types of inspections to be performed at different intervals. During your municipal fire inspections, your local fire inspector or fire marshal will want to see the documentation of these inspections. The record of these inspections is attached directly to the extinguisher in the form of stickers or hang tags.

You should have an inventory of all of the extinguishers in your building and a record of when each was placed in service. Don't forget to inventory and perform the necessary inspections to spare extinguishers not yet in service.

PM Tasks

Verify that extinguishers are in their proper locations and that there are no obstructions to their access. Check for obvious physical damage such as leaking, damaged hoses, or clogged nozzles. Make sure the instruction label is facing outward. Check any anti-tamper hardware for damage and check the pressure gauge for the proper pressure.

In addition to these tasks, there are other inspections and tests that must be performed by licensed contractors. These required inspections are listed in Figure 11-9.

Required Fire Extinguisher Inspections and Tests

Annual Maintenance	Inspection of condition by licensed contractor. Required for all extinguisher types.
5 year hydrostatic testing	Required for CO2 and wet chemical extinguishers.
6 year maintenance	Emptying and inspecting interiors of dry chemical and halon extinguishers
12 year maintenance	Required for dry chemical and halon extinguishers

* it is usually more cost effective to replace small portable extinguishers than to have 5,6, or 12 year maintenance performed.

Figure 11-9. Required fire extinguisher testing and inspection

Recommended PM Frequency

Monthly: visual inspection of condition and recording inspection date on attached tag.

Technical Notes

During annual service, your inspection company will attach an inspection tag to the collar of the extinguisher. This collar will have 12 boxes, one for each month, where a record of monthly inspections can be recorded. During fire inspectors, most inspectors will look on the back of these tags to see that annual visual inspections are being done.

Initial and date the tag each month when you complete the visual inspection.

See also Kitchen Fire Suppression Systems.

FIRE HYDRANTS

Fire hydrants provide a source of water for the fire department in the event of a fire. Most fire hydrants are located along public streets and are the property and maintenance responsibility of the local municipality. When fire hydrants are located on private property, maintenance may be the responsibility of the local municipality, or of the property owner. In many places, the local fire department or water department will help building owners perform these annual tests.

PM Tasks

NFPA 25 *Standard for* ***Inspection****, Testing, and Maintenance of Water-Based* ***Fire*** *Protection Systems* requires that fire hydrants are flow tested annually.

If your municipality allows this testing to be done in-house, you can do so by removing one of the fire department connection caps and opening the valve fully. Flow must be maintained for not less than 1 minute and until all debris have been cleared from the hydrant. After the flow test is complete, close the valve and make sure the hydrant barrel drains properly to prevent frost damage.

NFPA 25 Also requires that a flow test be performed every 5 years recording pressure and flow rate to verify the condition of the pipe supplying the

Figure 11-10. Fire hydrants should be flushed once each year.

hydrant. This test can usually be performed by your fire sprinkler service company if your local water for fire department cannot do so.

Recommended PM Frequency

Annually: Flow test

Every 5 years: Flow test with pressure and flow rate measurement.

Technical Notes

The valve seat of most fire hydrants is located either below the frost line for a ground hydrant or within the heated space of a building for a wall mounted hydrant. This prevents the hydrant from freezing in the winter. All hydrants are designed with weep holes to drain water out of the hydrant.

To remove the cap from the fire department connection or to open a fire hydrant (except for hydrants with hand wheels) you will need a hydrant wrench. A hydrant wrench is made to fit the 5-sided valve stem found on the top of the hydrant. Any other tool could damage the stem.

Whenever you are doing a flow test, be sure to notify the local fire department and water department. Flowing a fire hydrant can cause a drop in water pressure for other local water users and can dislodge dirt from municipal water lines causing dirty or cloudy water in the neighborhood for a few hours. Many municipalities perform flow tests on privately owned hydrants so they can open other hydrants in the area at the same time to wash this dirty water out of the pipes before neighbors complain.

FIRE INSPECTIONS (MUNICIPAL)

Most jurisdictions require commercial facilities to be inspected for compliance with fire safety codes. These inspections are often performed by a municipal fire inspector, municipal or county fire marshal or a member of the local fire department.

In most cases, the local authority having jurisdiction (AHJ) will schedule these inspections and will show up unannounced. In some cases, it may be the responsibility of each facility to call and schedule an inspection. Whether it is your responsibility to schedule an inspection or not, the inspection should still be on your PM calendar as a reminder that the inspection will be coming up and your facility needs to be ready.

Fire inspectors will be looking for compliance with local fire and building codes. These local codes usually reference most of the different codes and standards published by the NFPA including:

- NFPA 72 The National Fire Alarm Code
- NFPA 101 Life Safety Code
- NFPA 10 Standards for Portable fire Extinguishers
- NFPA 720 Standard for the Installation of Carbon Monoxide (CO) Warning Equipment in Dwelling Units
- NFPA 96 Standard for Ventilation Control and Fire Protection of Commercial Cooking Operations,
- NFPA 13 Standard for the Installation of Sprinkler Systems, NFPA 17 Standard for Dry Chemical Extinguishing Systems,
- NFPA 17A Standard for Wet Chemical Extinguishing Systems
- NFPA 110 Standard for Emergency and Standby Power Systems
- NFPA 25 Standard for **Inspection**, Testing, and Maintenance of Water-Based **Fire** Protection Systems.

PM Tasks

A building should always be in compliance with fire safety codes and should always be in inspection ready condition. However, when a fire inspection is pending, it's a good time to double check your building for compliance.

Your fire inspector will often want to see the following documentation:

- Inspection report for fire sprinkler system
- Inspection report for kitchen fire suppression system
- Inspection report for fire alarm system
- Record of cleaning for kitchen exhaust hood systems
- Monthly inspection tags on portable fire extinguishers
- Monthly inspection record for in-house testing of emergency and exit lights
- Monthly record of in-house fire alarm system testing.

All of these items are explained in more detail under their respective areas in this chapter.

Each fire inspector or department will have different inspection issues that they tend to concentrate on. The inspection focus list often changes from year to year. Some of the most commonly cited violations follow in no particular order:

- Extension cords may not be used as a substitute for permanent wiring. The NFPA considers any use over 90 days to be permanent. Note: Most jurisdictions will allow power strips to be used if they have built in circuit breakers. Surge suppressors are not the same as circuit breakers and are not a suitable substitute.
- Exit signs and emergency lights must be lit and the backup batteries must work.
- Electric space heaters cannot be used.
- Exit hardware must open in one motion. If you have to turn a key and then push the door to get out, that is a violation.
- Flammable materials cannot be stored in the building. This includes the gasoline in the fuel tank of a lawnmower or even oil-based paints. Flammable materials must be in a fire rated flammables storage cabinet.
- Combustible materials cannot be stored under stairways.
- No storage within 24 inches of the ceiling.
- No storage within 18 inches (vertically) of a sprinkler head.
- Pressurized cylinders (oxygen, helium, acetylene, etc.) must be secured from falling over.
- Fire lanes must be free of obstructions such as loading or unloading vehicles
- Combustible waste such as empty cardboard boxes cannot be stored in or near the building.
- Aisle ways, walkways or stairways must be kept free of obstructions.
- Electrical receptacle, junction box, or breaker panel covers must be in place.
- Smoke and fire separation doors must close and latch without assistance and seal properly.

Recommended PM Frequency

Most jurisdictions perform annual fire inspections. Some "high risk" industries such as hospitals, nursing homes, or schools may require more frequent inspections. Contact your local AHJ for information.

FIRE SPRINKLER SYSTEM

A building's water-based fire suppression must be maintained and inspected according to NFPA 25 *Standard for Inspection, Testing, and*

Maintenance of Water-Based Fire Protection Systems. Inspections and maintenance should be performed by qualified individuals, which usually means hiring an outside sprinkler company. In some jurisdictions, the people that inspect sprinkler systems will need to be licensed.

Make sure to maintain excellent records of all fire sprinkler inspections and repairs. Your local fire inspector or fire marshal will want to see these during your municipal fire inspections.

PM Tasks

Every part of the sprinkler system needs inspection or maintenance at different intervals. In wet pipe systems, pressure gauges are to be checked weekly to see that there is adequate water pressure and control valves are to be inspected weekly to make sure the valves are kept open. If control valves have tamper signaling devices, these only need to be inspected monthly. These visual inspections can usually be done by in-house staff if records are kept.

Fire pumps require weekly visual inspections in-house and annual testing. Dry system gauges should also be inspected weekly to verify that water has not entered the dry part of the system. Wet system gauges should be inspected monthly. These inspections can also be performed in-house.

A quarterly inspection by your inspection company should include testing alarm devices (flow switches), inspecting all fire department connections, and operating the main drain to verify that water will flow.

Finally, the entire system must be inspected annually including opening and internal inspection of any dry pipe valve, flow testing all parts of the system, inspection of the condition of all pipes, checking that spare sprinkler heads (and a head wrench) are available, testing of back flow preventers, operating all control valves through their full range of motion, and visually inspecting all sprinkler heads.

Recommended PM Frequency

Monthly (or Weekly): Check and record gauge pressures, inspect control valves in house.

Quarterly: Quarterly sprinkler inspection by sprinkler company.

Annually: Annual inspection by sprinkler company.

Technical Notes

Building fire sprinkler systems are usually one of two types. Both types may be present in some buildings. The more common of the two types is the wet system. Wet sprinkler systems have their pipes filled with pressurized water at all times. When a fire heats any sprinkler head to the factory set activation temperature (usually 135° to 170°F), that sprinkler head opens and water begins to flow to put out the fire.

The second common type of system is the dry system. Dry sprinkler systems use an air compressor to keep the sprinkler pipes filled with compressed air instead of water. Like a wet sprinkler system, individual sprinkler heads will open when the temperature reaches a pre-set point. When a sprinkler head opens, the air in the pipes is expelled. This drop in air pressure causes the main "dry pipe valve" to open; flooding the pipe with water. According to NFPA 13: Standard for the Installation of Sprinkler Systems, water must reach all points of the sprinkler system within 60 seconds of a sprinkler head activation. Dry systems are used in areas that are subject to freezing such as unheated attics or garages.

There are other types of sprinkler systems that are less common in most facilities. Hazardous locations such as airplane hangers or chemical storage may require deluge sprinkler systems in which all sprinkler heads flow water at the same time.

Pre-action sprinkler systems are used where there is a concern of water damage from a damaged or malfunctioning sprinkler head. A pre-action system uses automatic valves to hold back water until a heat or smoke sensor calls for water. Condominiums or townhouses sometimes use these systems with a separate valve for each living space. If a sprinkler head or pipe is damaged in a dwelling unit, no water will flow causing property damage. However, if a head activates from heat, the valve will open allowing water to flow, putting out any fire.

See also Kitchen Fire Suppression Systems, Fire Hydrants, Fire Extinguishers.

FIRE SUPPRESSION SYSTEMS

See Kitchen Fire Suppression Systems, Fire Extinguishers, Fire Sprinklers.

FLAG POLES

Flag poles are an often overlooked item and probably minor enough not to justify being included in this list if it weren't for the problem of trying to replace a broken rope. Once the rope has broken (and probably fallen to the ground), you'll need an aerial lift truck to get a new rope over the pulley at the top of the pole.

PM Tasks

Check the rope, replace the rope when damage is noticed by taping the new rope to the butt end of the old rope and pulling the new over the pulley using the old.

Recommended PM Frequency

Daily: Make note of the condition of the rope everyday when raising and lowering the flag.

Technical Notes

Since the flag is being raised and lowered each day, the condition of the rope can easily be checked at the same time. Plastic coated wire cable doesn't stretch like rope and lasts longer. Eventually the plastic coating begins to chip off and the cable will need to be replaced before the chipping causes the cable to bind in the pulley. If a rope does fall, many fire departments will assist with replacing the rope on a flag pole.

FLOOR CARE

See Carpet Care, Resilient Flooring Care.

FLOOR DRAINS

Like all sanitary drains, floor drains are constructed with plumbing traps to maintain a seal of water to keep sewer gasses out of the building. Many floor drains are installed in areas where they are never used and the water in the trap eventually evaporates letting sewer gasses in.

PM Tasks

Pour water into each drain to maintain the seal against sewer gas.

Recommended PM Frequency

Monthly

Technical Notes

Some floor drains have "trap primers" installed that automatically feed water to the drain to maintain the water level in the trap. These primed drains will not need regular PM unless the primer fails. Most primer valves are not repairable and are simply replaced if they no longer work.

A few drops of baby oil deposited in a floor drain will form a thin film of oil on the water's surface slowing the rate of evaporation and help to extend the PM interval.

Figure 11-11. Barrier type drain trap

There are also barrier type drain trap seals can easily be installed inside existing floor drains. These seals have a flexible rubber chute that opens to let water into the drain but closes to prevent sewer gasses from coming into the building. It can make sense to install these on difficult to access drains or on drains that seem to cause many complaints.

FURNACES

Forced air furnaces use electric heating coils or combustion of fuel (gas or oil) to directly warm air which is usually delivered to occupied spaces through ductwork.

PM Tasks

All of the combustion appliance checks that apply to boiler apply to furnaces. Combustion burners (oil or gas) should be cleaned and adjusted every year by someone qualified to do this work. Oil nozzles and oil filters should be replaced, oil pump pressure should be tested, and the combustion chamber should be cleaned and inspected. Gas appliances need to be cleaned and burners adjusted for the most efficient operation. Barometric dampers should be adjusted.

In addition, furnaces will need to have their heat exchangers inspected annually by someone qualified to determine if there are any cracks or holes.

Grease or oil bearings in blowers and motors, check the condition of belts and pulleys. Clean filters.

Recommended PM Frequency

Quarterly (during heating season): Inspect belts and pulleys, clean filters.

Annually (at start of heating season): Have furnaces professionally inspected and adjusted.

Technical Notes

A cracked heat exchanger will allow the gasses of combustion to mix with the air delivered to the building and there is a danger of carbon monoxide poisoning. Wherever combustion furnaces are used (or other combustion appliances) it is a good idea to install carbon monoxide (CO) detectors. CO detectors are often required in sleeping rooms when combustion appliances are present.

Also see "Boilers and Water Heaters" section of this chapter

GENERATORS

See Emergency Generators

GREASE TRAPS

Grease traps, also called grease interceptors or FOG interceptors (fat, oil, grease), are installed as part of a facilities waste water system to collect grease which can clog pipes. Grease traps are installed at locations where grease is most likely to end up in drains. These include dishwashers, pot sinks, and other drains in commercial kitchens.

Grease traps can be small rectangular tanks installed in a kitchen floor or can be larger tanks with man hole covers installed outside the facility. Grease traps work by using baffles that block floating grease from leaving the tank and by having a discharge pipe that is below the level of floating grease.

PM Tasks

Accumulated grease needs to be removed when traps are approximately 25% filled with grease. With in-floor grease traps, this involves

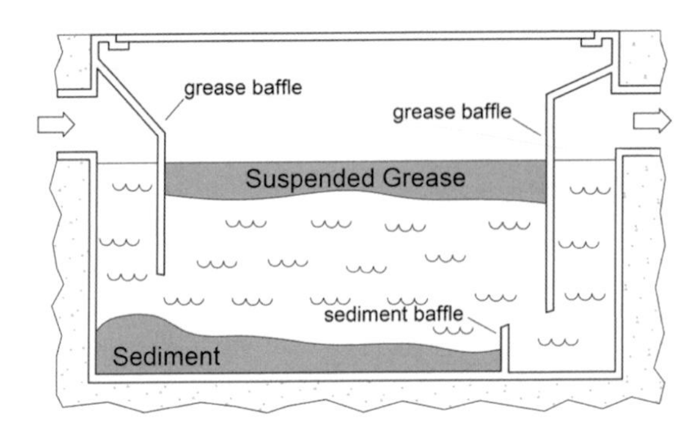

Figure 11-12. Components of a grease trap

removing the grease trap cover and scooping the grease out of the trap into a bucket. For large grease tanks, a septic tank pumping service can be contracted to remove the grease.

Recommended PM Frequency

Quarterly: Cleaning grease traps quarterly is probably a good starting point. Kitchen usage and the type of cooking done will determine the required frequency which should be adjusted as needed to keep traps from becoming more than 25% full.

Technical Notes

Most companies that provide janitorial chemicals will offer enzymes which break down grease and extend the time between grease trap cleaning. Usually, a small chemical pump is installed at the sink which discharges into the grease trap. A timer dispenses a small amount of the enzyme into the grease trap after-hours so the enzyme has time to work on the grease in the trap.

HOT WATER HEATERS

See boilers and hot water heaters

HOT WATER TEMPERATURES

Excessive water temperatures at sinks, bath tubs, showers, or other locations can cause scalding. Children under 5 and the elderly are most susceptible to burns from hot water but hot water can burn anyone. Most adults will suffer third-degree burns if exposed to 150 degree water for two seconds; 140 degree water for six-seconds; or 130 degree water for 30 seconds. Even at water temperatures of 120 degrees, third degree burns can occur with an exposure of 5 minutes.

Facilities such as schools, hospitals, nursing homes, assisted living facilities, and day care centers are often subject to regulations concerning potable water temperatures. In all types of facilities, excessively hot water is a safety and liability issue.

Under no circumstances should hot water exceed 120 degrees.

PM Tasks

Hot water temperatures should be taken monthly unless other regulations apply. Hospitals are often required to take water temperature readings daily and to keep a permanent record of the measurements.

Temperature readings should be taken in each area that is served by a different source of hot water and should be taken in a different location each month. The easiest way to measure the hot water is to place a drinking cup in a sink, insert a thermometer into the cup, and run the hot water into the cup allowing it to continue to overflow into the sink until a steady temperature is reached. If the water temperature exceeds 120 degrees at any time during this test, adjustments must be made.

Recommended PM Frequency

Monthly: Measure the water temperature in each area served by a different source of hot water. More frequent measurements may be required for specific industries.

Technical Notes

If water temperatures are too high, adjustments should be made to hot water heater thermostats, or to mixing valves if they are used. Mixing valves are installed as a way to extend the supply of hot water. Hot water at high temperature (often 140-160 degrees) and cold water are supplied to the inlet ports of the mixing valve. Inside the mixing valve, the hot and cold water is mixed provide water at the desired temperature. Mixing valves automatically and continuously adjust the mixture of hot and cold to maintain this desired outlet temperature. When measuring water temperatures, the area served by each mixing valve must be tested individually.

Small mixing valves can be installed directly at each sink or other fixture. Some faucets and showers have mixing valves built-in. If your facility has any of these types of mixing valves or fixtures, you will need to measure water temperatures at each and every fixture because mixing valves can fail.

Anti-scald faucets or anti-scald shower heads are excellent ways to prevent anyone from getting burned. If water temperatures exceed 120 degrees, these devices will automatically shut off the flow of water. Anti-scald devices may be required in buildings such as schools and healthcare facilities.

HVAC

See Air Conditioners, Chillers, Cooling Towers.

ICE MACHINES

Ice machines will run more efficiently and last longer if they are maintained properly. Ice machines should also have a regular cleaning schedule to prevent the spread of diseases.

PM Tasks

Clean evaporator coil, pump, and water reservoir. Change condenser coil filter and check to make sure the condenser coil is clean. Drain and clean the ice storage bin with a sanitizing cleanser. Check the operation of augers. Test ice level sensor to make sure it will shut down unit when bin is full.

Recommended PM Frequency

Monthly: Clean and inspect.
Annually: Change water filter.

Technical Notes

Evaporator coils are often coated with a layer of nickel or other material to help the ice cubes release easily. Using the wrong cleaning chemicals can damage this surface and ruin the evaporator coil. Use the cleaning chemicals recommended by the manufacturer.

IRRIGATION

See "Lawn Irrigation."

KITCHEN EXHAUST HOODS AND FANS

This section applies to exhaust hoods that are located directly over cooking appliances in commercial food service kitchens. NFPA 96 refers to these as "grease removal exhaust fans" and prescribes specific preventive maintenance tasks that must be performed. Kitchen exhaust fans that are only for the removal of hot air or steam should be maintained as in the section titled "Exhaust Fans."

Commercial food service kitchens are required to have a grease removal exhaust fans and hood assembly located over grill tops, fryers, and other cooking equipment. Most of these hoods will include a fire suppression system with spray heads located directly over each piece of equipment.

PM Tasks

Grease filters, located directly over cooking appliances, should be cleaned as often as necessary to prevent grease build-up. Depending on the type and volume of cooking done, this could mean monthly cleaning or even running the filters through the dishwasher at the end of

Figure 11-13. Kitchen exhaust hood

every night. This is a maintenance item usually handled by the kitchen manager, not maintenance staff.

NFPA 96 requires that the hood system, fan, and ductwork be inspected and cleaned prior to becoming heavily contaminated with grease. This will also vary with the type of cooking done. NFPA offers the following guidelines: hoods used for solid fuel cooking such as wood and charcoal should be cleaned monthly. Charcoal, char broiler, wok, and 24-hour cooking establishments should have the hood cleaned quarterly. Moderate volume kitchens need cleaning semi-annually. Seasonal and occasional cooking such as camps, churches, and senior centers only require annual cleaning. Remember that these are guidelines and the amount of grease deposit is what matters in determining when it's time to have the hoods cleaned.

Exhaust hoods must be cleaned by someone that is qualified, trained, and certified to do the work. Since NFPA does not offer certification, most "qualified" service people are certified by one of several private certifying organizations. The largest and most often cited is IKECA, the International Kitchen Exhaust Cleaning Association.

Recommended PM Frequency

Monthly, Quarterly, Semi-annually, or Annually. Depends heavily on the type and volume of cooking done. See the above section for details.

Technical Notes

The actual cleaning of your exhaust hood involves creating a water collection system below the exhaust hood using plastic sheeting and pressure washing (or steam cleaning) every internal surface of the ductwork and exhaust hood. All duct bends must have a door for access to the interior and there must be access doors at least every 12′ to allow cleaning personnel to see and clean every part of the interior.

The exhaust fan will have its housing or cover removed and it too will be pressure washed. The job is complete when the entire ductwork system is cleaned to bare metal.

Also see Kitchen Fire Suppression Systems, Exhaust Fans

KITCHEN FIRE SUPPRESSION SYSTEMS

Because of the presence of natural gas and cooking oil, fires that occur in and around cooking equipment in commercial food service kitchens often spread rapidly. Almost every commercial kitchen has a kitchen fire suppression system installed in the grease exhaust hood over cooking appliances.

Commercial kitchen fire suppression systems from all manufacturers are often incorrectly referred to by facilities people simply as Ansul® systems. Ansul® is a subsidiary of Tyco Fire and Safety is one of the largest manufacturers of fire suppression equipment.

PM Tasks

Each month the system should be visually inspected. This can be done by in-house staff. Check the tank pressure gauge for adequate pressure, look for missing grease caps on spray nozzles, check that nozzles are in their correct locations and not damaged or blocked. Make sure any manual activation stations have the safety pin in place and are properly labeled. Look the system over for any obvious signs of problems. An inspection tag located on the chemical tank should be initialed monthly as a record of the inspection.

There are several different regulations that apply to kitchen fire suppression systems depending on the type of suppressant used. NFPA 96 applies to grease removal exhaust hoods, NFPA 13 to water-based fire suppression systems, NFPA 17 to dry chemical systems, and NFPA 17A to wet chemical systems. All of these standards require that inspection and servicing of the fire-extinguishing system shall be made at least

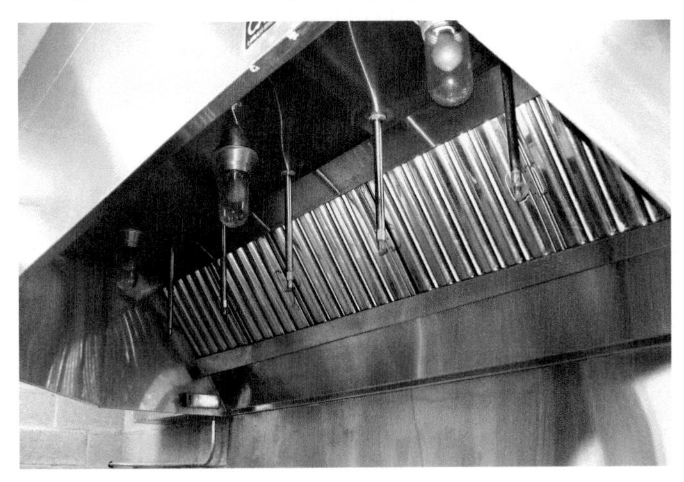

Figure 11-14. Spray nozzles of a kitchen fire suppression system are located under the exhaust hood.

every 6 months. In most jurisdictions inspections must be made by a certified or licensed inspector or installer.

Recommended PM Frequency

Monthly: Visual inspection by in-house staff.

Semi-annually: Inspection and testing by certified contractor.

Technical Notes

Kitchen fire suppression systems consist of a set of spray nozzles located over each cooking appliance, fusible links (which melt in the heat of a fire) or electronic heat detectors to activate the system, and a pressurized tank of fire suppressant which is sprayed from the nozzles to put out the fire. When the system activates, any kitchen makeup air fans (but not exhaust fans) and the gas supply to the cooking appliances must shut down automatically.

Due to the change many restaurants and kitchens have made toward healthier but hotter burning cooking oils, older dry fire suppression systems are now considered obsolete. Many food service kitchens are being required by insurance companies or municipal inspectors to upgrade to newer wet chemical systems.

During your semi-annual inspection and test, the inspecting contractor will leave a dated inspection tag attached to the suppression chemical tank. Your local fire marshal or fire code inspector will want to see this tag during your municipal fire inspections.

See also Kitchen Exhaust Hoods and Fans, Fire Extinguishers, Fire Sprinkler Systems.

LAUNDRY (COMMERCIAL WASHERS AND DRYERS)

Hotels, hospitals, and nursing homes usually have their own in-house commercial laundry operation. Most commercial laundries include conventional washers and dryers. The largest laundry operations can include folding machinery and often employ full time laundry mechanics to maintain this equipment. Most of us will only be maintaining washers and dryers.

PM Tasks

Dryers: Inspect drive belts for condition and tightness. Grease bearings. Clean burner assembly of any lint. Clean lint from machine. Inspect inside of exhaust ducting for excessive lint or blockages.

Washers: Check hoses for signs of wear or deterioration. Check condition of door seals. Inspect drive belts and grease bearings. Check under machines for signs of leaks. Check operation of chemical pumps and verify with laundry staff that machines are not consuming chemicals at an unusual rate.

Recommended PM Frequency

Monthly

Technical Notes

Inspections should be done monthly, greasing of bearings may need to be done less or more frequently, consult the manufacturer's manual.

LAWN IRRIGATION

In almost all locations, the only way for a facility to have a lush, green lawn throughout the spring, summer, and fall is through mechanical irrigation. Typically lawns require 1" to 1¼" of water each week. Two or three deep waterings each week will be much more beneficial than shallow watering every day.

One of the drawbacks of irrigation is grass that is often wet is susceptible to fungus damage, usually noticed as brown patches in the lawn. The best way to prevent fungus is to water in the early morning instead of evening. This allows the water to soak into the ground but lets the blades of grass dry quickly in the morning sun.

PM Tasks

Season Start-up: At the start of the irrigation season, control timers need to be programmed and their batteries changed. Each irrigation zone and each and every head should be checked for operation. Most service companies will verify that each head pops up and sprays water. Few will take the time to watch each head to be sure it moves through its full range of motion. This is a time consuming but important part of making sure the system is doing its job.

Backflow prevention devices are required on irrigation systems. Since pesticides and fertilizer can be present in the lawn, these chemicals could be siphoned back into the potable water supply if the water supply was to lose pressure. Back flow prevention devices will prevent this from happening. Backflow devices must be inspected by a certified tester or licensed plumber once or twice per year. Most jurisdictions only require one annual test for irrigation systems since they are only used for half of the year. See "Backflow Prevention Devices" for more information.

After initial startup, the best PM is to walk the lawn once a week and look for areas that are not getting watered. Pop up heads can be damaged by vandalism or mowers. Irrigation pipe underground can break causing washed out holes in the lawn. Spray heads can become clogged by dirt which may prevent them from spraying or can jam the rotation mechanism. Sprinkler valves can fail to open or fail to close. Control timers can lose their programs or become damaged by power surges or lightening. All of these problems can be noticed by a weekly walk of the lawn. During a hot dry summer, a lot of lawn damage can be done in a few days if the irrigation is not working.

Seasonal Shutdown: At the end of the irrigation season, timers should be turned off, the water supply to the system should be turned off and the irrigation system blown out with compressed air to remove water that could freeze over the winter and cause damage. An air compressor is hooked to an air fitting near the main shut off valve and each zone is operated until only a light mist of water is seen coming from the sprinkler heads at each zone.

Recommended PM Frequency

Annually: System start up in the spring including backflow prevention device test.

Annually: System shut down in the fall.

Weekly: Walk lawn and look for any signs of a system, zones, or individual heads that do not work.

Technical Notes

Lawn irrigation systems include various types of sprinkler, mist, or spray heads which are fed water by a network of underground pipes. Most irrigation heads sit flush or slightly below the level of the ground and are pushed up above the surface by water pressure. Electrically operated valves called "solenoid valves" turn on the different irrigation zones. Electric wires buried with the underground pipe carries the electricity to operate these valves from the control timer. Sprinkler systems typically operate at 24V.

Brown or dead triangles, or pie wedges, of grass indicate that a sprinkler head is not rotating its full range of motion. Green growth in circles around the spray heads indicate a drop in water pressure has resulted in a smaller spray circle. This is usually caused by two zones operating at the same time, a partially closed valve, or a failed booster pump. Larger browning patches of grass usually indicate a bad zone, either due to a programming problem, a bad wire to a zone valve, or a failed zone valve.

LAWNMOWERS AND GROUNDS CARE EQUIPMENT

Lawnmowers, leaf blowers, garden tractors, chainsaws, hedge trimmers, and other grounds care equipment all need to be included on your PM calendar or schedule.

PM Tasks

Review the manufacturer operator's manuals for the specifics for each piece of equipment as they all vary widely. Typically, larger equipment will need oil changes every 50, 100, or 200 hours of operation, grease fittings will need to be greased, mower blades kept sharp and balanced, trimmer blades and chains oiled, belts inspected for wear, and tire pressure maintained. There will be other items required depending on the type of equipment you are using. In all instances, the manufacturer's operating manual will explain exactly what type of maintenance is required for each type of equipment.

Since this type of equipment is usually maintained based on hours of operation, and not based on a calendar date, it makes sense to affix a service label to each piece of equipment that says "Next service at _____ hours." For equipment without an hour meter, you can estimate based on number of hours used in a typical day or you can buy inexpensive after-market hour meters.

Recommended PM Frequency

Depends entirely on the type of equipment. See the manufacturer's literature.

LIGHTING (EXTERIOR)

Exterior lighting is clearly a safety issue as well as affecting the overall appearance of your facility.

PM Tasks

Turn on timers or cover photo-eyes to cause lighting to come on. Walk the exterior of the facility and take notes on any fixtures that are not working and make repairs.

Also consider group re-lamping for HID fixtures as discussed in Chapter 1.

Recommended PM Frequency

Monthly: Walk facility in the evening to check all exterior lighting.

Quarterly: Adjust lighting timers for seasonal daylight changes.

Technical Notes

Many types of high intensity discharge lights (HID) will turn themselves off and on all night when bulbs start to go bad. A light that is on when lamps first light may turn itself off within 10 or 15 minutes. Lights that turn on and off are a sign of a lamp going bad. A HID light that fails to light will most likely have a bad bulb but may have a bad ballast.

Changing pole lights is easiest if done using an aerial lift truck or boom lift. It may be possible to change some pole lights using extension ladders. If you don't own an aerial lift, you can rent them from many places. It may be more cost effective to have a standing agreement with your electrical contractor to replace bulbs or ballasts that you stock each month.

HID lighting has a trait known as lumen maintenance. This is a measurement of how well a light maintains its light output over the life of the lamp. HID lamps loose intensity as they age and can loose as much as 40% of their light output near the end of their life. It often makes sense to replace HID lamps once they reach a certain percentage of their expected life or a specific loss in lumen output. See the section on group re-lamping in Chapter 1 for more details.

LIGHTING (INTERIOR)

Interior lighting should be checked in each area of the building when you are doing Room PM (See section on "Room PM" in this chapter). All areas will need to have their lighting checked as part of room PM or separately.

PM Tasks

Operate lights and verify operation.

Recommended PM Frequency

Monthly. Consider group re-lamping of HID or fluorescent lighting, especially in difficult to access areas such as high ceilings. See Chapter 1 for information on group re-lamping.

LOCKS

Conventional key-in-knob type locksets require very little preventive maintenance. Most commercial locksets are classified as either Grade 1 or 2 by the Builders Hardware Manufacturer's Association (BHMA). Grade 2 locks are tested to function for 400,000 cycles and grade 1 locks are required to survive 800,000 cycles during testing. Many lock manufacturers produce locks that exceed these requirements and can be cycled into the millions of times without any malfunctions.

Electronic locks will usually require more frequent PM than conventional locks.

PM Tasks

For conventional locks, the only maintenance that is necessary is to make sure that the locks screws have not become loose which requires occasionally tightening the screws and to occasionally lubricate the keyway. Lock keyways should only be lubricated with dry graphite lubricant. Any oil in the delicate mechanism of the keyway would attract dirt and could jam the lock cylinder mechanism.

Electronic locks provide many useful features such as the security of being able to remove a key card or code from the system without having to mechanically re-key any locks. They can also provide an audit trail of when, where, and who had access to any part of your facility.

Many electronic locks require more frequent maintenance than traditional key locks. The maintenance requirements vary greatly from system to system but often include scheduled battery changes and cleaning magnetic heads in card reader systems. See your lock manufacturer's literature for the PM requirements for your particular system.

Recommended PM Frequency

Annually: Inspect locks, tighten screws if necessary, lubricate keyway. Electronic locks will have specific maintenance requirements specified by the manufacturer.

LOCKERS

Schools spend a considerable amount of time each summer preparing student lockers for the following school year.

PM Tasks

Change lock combinations. Clean graffiti. Remove any mirrors, dry erase boards, photos left by students. Inspect lockers for operation and make repairs.

Recommended PM Frequency

Annually: Repair lockers.

Every 5 years: Have lockers professionally repainted.

Technical Notes

Locker PM is best handled similarly to room PM. A PM cart with all necessary replacement locker parts and tools should be prepared so that your maintenance person can move this rolling shop from locker to locker solving each problem as it is found. Locker parts are available from several sources.

Most lockers have combination locks that can be changed each year. These locks have a factory set sequence and a change key is used to advance to the next combination. Careful records of each locker's combination should be kept including any locks that are replaced and the new lock's factory combination sequence.

NURSE CALL SYSTEM

If you work in a hospital or nursing home, you are familiar with nurse call systems. These allow patients to call for assistance by pushing a button on a handheld pendent, bed mounted panel, or other device. Nurse call systems ring an alarm, send a page to a nurse's pager, or in some way notify nursing staff that a particular patient needs assistance.

Nurse call systems and the nurse's response time is a component of state, federal, and JCAHO, facility inspections. Nurse call systems should be considered life-safety equipment and as such should have a high priority in you PM program.

PM Tasks

Test each nurse call station to make sure that the nurse is notified of the call, that any intercom operates correctly, and that the nurse call indicator lamp outside the patient room lights.

Recommended PM Frequency

Monthly.

Technical Notes

Since this is a life-safety issue and subject to inspection by so many agencies, be sure to maintain excellent documentation of inspections and repairs.

PARKING LOTS AND SIDEWALKS

PM Tasks

Inspect parking lot for pot holes, standing water, alligatoring (cracking), or tire ruts that form at turns. Pick up broken glass and other trash. Inspect the condition of signs. Outside contractors can be hired to sweep using street sweeping vehicles at a lower labor cost than doing so with broom and trashcan. Some companies apply liquid sealer to the asphalt surface annually to extend asphalt life. Re-striping should be included in the budget roughly every 5 years depending on the level of traffic and aesthetic standards.

Sidewalks need to be inspected for uneven surfaces, or any cracks that could cause a person to trip or fall. Any change in elevation over ¼″ should be considered a tripping hazard.

Recommended PM Frequency

Monthly: inspection of parking lot and sidewalks
Semi-annually: street sweeping
Annually: seal coat with asphalt
Every 5 years: re-stripe parking spaces and traffic lanes

Technical Notes

When filling potholes, all loose asphalt must first be removed from the hole. Asphalt patch should be tamped firmly. Tamping tools can be dipped in kerosene to prevent the asphalt from sticking. In cold weather, vehicle tires can be used to tamp stiff asphalt.

Holes over 2′ in diameter or large areas of cracked and spalling asphalt will require professional hot patching. Damaged areas should be cut back to good asphalt using an saw and the area filled with hot asphalt.

Water penetrating into asphalt is the biggest cause of parking lot damage. Having repairs made to low areas that tend to pond and making sure to seal cracks can extend the life of a parking lot by several years. Annual seal coating helps to keep water on top of the surface. Parking lots can typically be expected to last 30 to 40 years with one major re-surfacing project at approximately the middle of the parking lot's useful life.

Areas where vehicles come to a stop or make a turn are subject to stress and tend to become damaged more quickly.

PLAYGROUND EQUIPMENT

The playground safety standard most often used in the US is the *Handbook for Public Playground Safety* published by the U.S. Consumer Product Safety Commission. This 47 Page guide is available free from the CPSC in print form or can be downloaded from their website. This book is often referred to as the playground bible and should be on every maintenance manager's book shelf.

Playground inspections can be completed in house or by a certified playground safety inspector (CPSI). Some schools or other facilities with playgrounds choose to have their in-house staff certified by attending a 3 day training course and passing the CPSI certification exam offered by the *National Playground Safety Institute* of the *National Recreation and Park Association*. Many organizations perform in-house inspections quarterly and have a certified inspector make inspections annually or every two years. Your insurance carrier may be able to provide playground inspections at no cost to you.

PM Tasks

All playground equipment should be inspected for corrosion, rot, insect damage, splintering, or weathering. Pay careful attention to moving parts where wear is likely to occur and fasteners that could become loose. Check for sharp points, corners, and sharp edges. Inspect the playground area for broken glass or other hazards. Surfacing material such as wood chips or shredded tires should be inspected to make sure they have not become compacted or displaced from high traffic areas. The *Handbook for Public Playground Safety* includes an excellent checklist that can be used as a guide to inspections and as documentation that regular inspections have been completed.

Since surfacing material tends to compact, decay away, or be distributed away from fall zones, new surfacing material is typically added each year.

Recommended PM Frequency

Quarterly: Inspection of all playground equipment using CPSC checklist.

Annually: Add surfacing material.

Technical Notes

Surfacing material is required to be deep enough to prevent head injuries in the event of a fall. This is typically 6-12" of material. See the Handbook for Public Playground Safety for specifics of surfacing depth and fall zones for different types of equipment.

Playground surfacing material is probably the part of playground maintenance that will require the most time and labor. Wood chips tend to compress and all loose fill materials tend to become displaced from high traffic areas and will need to be raked back into place or more material may need to be added on a regular basis. Depending on the amount of playground usage, this may need to be completed daily.

POOLS

See Swimming Pools.

PUMPS

See Circulator Pumps.

RADON MITIGATION EQUIPMENT

Radon is a naturally occurring radioactive gas that seeps from the ground into buildings in some areas. Radon ventilation systems use exhaust fans to draw air from under a building and vent that air to the outside.

See also Exhaust Fans.

RAIN GUTTERS

Rain gutters collect rain water from pitched roofs. Like roof drains, rain gutters need to be kept clear of debris. Leaves, dislodged roofing mineral material, and foreign objects can block gutters. Because clogged gutters retain water, any dirt that collects in gutters is an excellent place for weeds, and even trees to grow.

PM Tasks

The one way to prolong the life of a gutter system is to keep it free of dirt and debris. Clogged gutters, or debris that slow water flow cause standing water. Standing water freezes and separates joints. Standing water accelerates rust and corrosion. Standing water can damage several types of sealants and caulks.

PM of gutters involves cleaning the gutters frequently. Depending on the circumstances, this can be done from the roof, on a ladder, from

Figure 11-15. Besides allowing water to enter the building envelope, clogged gutters will rust or corrode and can support plant growth.

scaffolding, or from aerial lift equipment. Gutter and leaders (the portion that runs vertically to the ground) need to be cleared of debris and flushed clean.

Open joints, missing end caps, and separated leaders need to be repaired immediately or water can enter the building envelope causing damage.

Recommended PM Frequency

Semi-annually: Clean gutters. It is best to schedule one gutter cleaning in the late fall after the leaves have fallen.

Technical Notes

Several different devices are available to prevent leaves and other debris from getting into gutters. Most of these are not marketed toward commercial properties and will not fit commercial gutter profiles. Basket shaped screens area available that prevent debris from entering leaders.

REFRIGERATORS AND FREEZERS

Like air-conditioners, commercial reach-in and walk-in refrigerators and freezers will benefit enormously from regular PM. Efficiency, the ability to hold temperature, and the life of the equipment will be affected.

PM Tasks

Clean and degrease condenser coils. Change or wash filters. Oil condenser fan motors. Check the condition of door gaskets and hydraulic door closers. All walk in boxes must be able to be opened from the inside, even with the door locked. Check interior lighting. Check evaporator coil for ice build up. Check freezer ceilings for ice stalactites indicting that de-icing heat coils are creating steam. Check de-icing timers. Check thermostat settings and box temperatures.

Recommended PM Frequency

Quarterly

Technical Notes

For more information please see Chapter 7—HVAC systems.

RESILIENT FLOORING CARE

Resilient flooring includes vinyl composition tile (VCT), vinyl tile and sheet flooring, linoleum tile and sheet flooring, and rubber tile and sheet flooring. VCT makes up the majority of resilient flooring in commercial buildings.

PM Tasks

Floor care starts with a sealer and several coats of floor finish. This floor finish should be dust mopped and wet mopped frequently to prevent dirt and other materials from becoming embedded in the finish. Spray buffing with a rotary floor buffer will help to maintain a floor's shine. Floors can also be cleaned with a soft rotary pad and then extra layers of floor finish can be added.

Once a floor has reached the point that buffing no longer is able to do an adequate job of maintaining the appearance of the floor, it is time to use chemical strippers to completely remove the old finish and start over with a sealer and several new layers of mopped on floor finish.

Recommended PM Frequency

The PM schedule for resilient flooring care will vary with buildings. Schools will usually be mopped each night but only stripped and refinished during the summer or possibly summer and winter break. Hospitals may refinish floors monthly. Office buildings may only require mopping weekly and refinishing annually. The frequency of these tasks must be determined for each building individually. The following information will need to be adjusted for your building.

Daily: mop floors.

Monthly: spray buff with rotary floor buffer or surface clean with rotary pad and add coats of finish.

Semi-annually: strip all finish from floors, seal, and add at least 3 layers of finish.

Technical Notes

Dirt works as an abrasive to remove the finish from resilient flooring as well as becoming embedded in the finish material. The best way to keep resilient floors looking their best is to prevent dirt from entering your building. See "Carpet Care" for information on using walk-off mats to keep dirt out.

There are many different systems of cleaning, stripping, sealing,

Figure 11-16. Shine of a well maintained vinyl composition tile (VCT) floor.

and finishing chemicals available for resilient floors. You can get systems created to provide high gloss or systems that promise low gloss to hide imperfections. Some finishes are long lasting but require high-speed equipment while others are easy to apply and maintain but may require more frequent reapplication. Your janitorial supplier will be able to help you sort through the options to find what works best for you.

Low-speed buffers are often used for stripping. High-speed buffers create a shiny, gloss surface. Exceptionally fast buffing machines, called burnishers are able to create a gloss surface on exceptionally hard finish materials. These machines are often used in shopping malls and department stores to create hard surfaces that can hold up to heavy foot traffic and still maintain a shine.

RETENTION PONDS

Retention ponds and detention ponds are both types of man made ponds intended to catch storm water. In many developed areas, fields, forests and other places where rainwater would be absorbed into the soil naturally have been lost to development. Without these natural watersheds, storm water runoff can cause streams and rivers to overflow. Man made retention and detention ponds provide temporary storage for these waters.

PM Tasks

Retention and detention ponds need little maintenance other than to prevent vegetation overgrowth from clogging the basin. Grass or other small vegetation is needed to prevent erosion from flowing water but retention ponds can quickly become overgrown with trees and brush. If your facilities lawn care is performed by an outside contractor, you should consider having maintenance of your retention pond written into the contract.

Maintaining retention ponds involves inspecting the pond for clogged discharge or inlet culverts. Making sure that any large rocks placed for erosion control are not moved from where they are needed, and keeping the basin area free of large vegetation by mowing, or by clearing the basin with other equipment.

Recommended PM Frequency

Quarterly: Inspect retention pond.
Annually: Cut and remove brush.

Technical Notes

Retention ponds hold storm water until the water is absorbed into the surrounding soil while detention ponds hold storm water only temporarily and slowly discharge storm water into streams or wetlands at a flow rate that will not overwhelm the watershed. Retention ponds do not have outlets while detention ponds do.

In addition to managing the volume of storm water released to a watershed, detention ponds also protect our streams, rivers, and wetlands from contamination. Surface contaminants carried to the detention basin have time to settle as sediment and are not carried away from the local area.

ROOFS ☎

Please read Chapter 6 on commercial roofs. Roofing PM is one of the most beneficial areas of preventive maintenance if done correctly and can be one of the most expensive if done incorrectly. Roofing PM is important enough that you can find an entire section in this book on the technical aspects of roofing PM in Chapter 6. Make sure to read that section when setting up your PM system or when performing roofing PM.

PM Tasks

Visual inspections for penetrations, damaged flashing, delaminated roofing fabric, rust, ponding, clogged roof drains and scuppers. Any debris should be removed from the roof to prevent puncture damage. Thermal imaging can be used to determine if roof insulation has become water saturated.

Recommended PM Frequency

Monthly: Inspect roof for any signs of damage according to inspection checklist in Chapter 6.

Annually: Contracted inspection by the installing contractor during the warranty period

Every 5 years and prior to end of warranty: Have independent 3rd party inspect roof for any defects including thermal imaging for

Figure 11-17. Commercial roofs have many different materials and components that need regular inspection.

saturated insulation. Apply seal coat to roof surface according to roofing warranty and recommendations of roof consultant.

Technical Notes

Please see Chapter 6 Commercial Roofs.

ROOM PM

In addition to performing PM to the mechanical equipment in a building, it is also important to perform PM to rooms within the building. The term "Room PM" comes from the hotel industry where a large percentage of PM time is spent making sure each and every guest room is in perfect condition. Room PM is not just for hotel guest rooms and should be performed in all parts of your building and in every type of building. Room PM should be performed in offices, the main lobby, rest rooms, employee break rooms, conference rooms, computer rooms, classrooms, dining rooms, patient exam rooms, and every other type of room found in your building. Common areas such as corridors and entrance foyers that would not exactly be considered rooms still need preventive maintenance and should be included in the schedule of Room PM.

Room PM will probably show up on your PM calendar more than all of the other PM tasks you will do combined. Only a few rooms can be checked each day so you may have Room PM on your schedule dozens of times before you have completed every part of your building. Several rooms can typically be maintained each day so your schedule may list rooms 101 through 107 for Monday; Rooms 108 through 114 for Tuesday; and the conference room, employee break room, and sales offices for Wednesday.

Room PM isn't about inspecting individual pieces of equipment. Room PM requires inspection and repair of every item within a room. The easiest way to do this is to create a PM cart. This is a rolling cart filled with all of the tools, spare parts, fasteners, caulk, paint, and whatever else is needed to solve all of the problems you are likely to find in a room. There are so many different items involved in Room PM that using a well stocked PM cart is the best way to avoid dozens of trips back to the shop or supply room for tools and parts.

PM Tasks

Room PM is usually done using a long checklist. No checklist can cover everything you might find in every type of room and a checklist still leaves open the possibility that something might be forgotten. An effective way to be sure that you are not missing anything is to do your inspection by working your way around the room and use your checklist after you are done as a double check.

To perform a thorough inspection, pick a location on the edge of the door to start your inspection, often next to the entrance door. Then continue around the room either to the right or left. As you move around the room, check each building component or piece of equipment that you come to. Next to the door you might first come to a light switch. Operate the switch and make sure the light works. To the left of the switch you may find a wallpaper seam that is separated. Note that on your work list. As you continue moving to the left you might next come to a closet door. Check the lock, hinges, door finish, and make sure the door opens and close properly. Continue around the room, checking every item, until you are back at the door where you started. Wallpaper, carpet, ceilings, bathroom fixtures, towel bars, televisions and appliances, furniture, electrical outlets, air conditioning units, and all other items in the room should be checked.

After the initial inspection, you should complete the work on the work list you just created and then move onto the next room on your PM schedule.

Recommended PM Frequency

The more often you do PM, the easier PM will be each time. If you are doing monthly PM to hotel rooms, the entire process could take as little as 10 minutes per room. If you only do PM quarterly, you will find yourself in the room for much longer.

The frequency of room PM depends on company standards, how often a room is use, and who uses the room. Hotel guest rooms, hospital patient rooms, and other rooms that get used often and where people have lots of time to inspect their surroundings should be maintained more frequently. Main lobbies, board rooms, and other areas that should impress the public should also be on the list of areas getting frequent room PM.

Offices, employee break rooms, and other areas rarely seen by the public do not need preventive maintenance as often.

As a guideline: PM high visibility areas monthly and other areas quarterly. Adjust these to fit your facility.

Technical Notes

Part of room PM involves aesthetic issues. Items such as smudged paint, dirty wallpaper, carpet stains, or scratched furniture are sometimes out of the thought process of skilled trades people used to performing more technical maintenance duties. Although some repairs found during room PM will require technical skills, room PM is entirely different than the work most maintenance people are used to.

Even skilled and experienced trades people may not be familiar with many of the tricks of the trade of room PM. Wood colored markers can quickly fix furniture scratches. Miniature paint sprayers filled with the right paint can do quick and neat touchup painting. Colored caulk can hide shrinkage gaps in wallpaper seams.

SEPTIC TANKS

In areas not served by local sewer, each facility will have its own system of dealing with plumbing waste water. In most cases, this will include a leaching field to dispose of liquid waste and a septic tank to store solid waste. Septic tanks need to be pumped out by a mobile septic service when they become full.

PM Tasks

It's a good idea to have septic tanks pumped before they become full. A full tank will cause plumbing drains to flow slowly or to backup and an over-full septic tank will be an odor problem.

After being pumped, your service company should perform a tank inspection.

Recommended PM Frequency

Your facilities history will be the best indicator of how often this service will need to be performed. If in doubt, it is a simple task to lift the concrete tank cover to see inside. Depending on the size of the tank, the size of the facility, and usage, tanks may need pumping several times a year or may not need to be pumped for several years.

Technical Notes

Septic service companies are often required to be licensed by the state or county.

SIDEWALKS

See Parking Lots and Sidewalks.

SMOKE DETECTORS (BATTERY OPERATED)

Multi family residences, hotels, and other buildings with individual sleeping rooms or apartments will often have single station battery operated smoke detectors. Just like the smoke detectors that are part of a building's fire alarm system, these detectors need to be tested monthly. Smoke detector batteries also need to be changed on a regular schedule. Fire investigations show that 25% of homes have smoke detectors that do not work. The primary reason is dead or disconnected batteries.

PM Tasks

Test smoke detectors by pressing the test button. Make sure that any interconnected smoke detectors sound the other detectors on the same circuit (usually in the same apartment or hotel suite). Change batteries twice each year.

Recommended PM Frequency

Monthly: Test smoke detectors

Semi-annually: Usually done during October (National Fire Safety Month) and April each year.

Technical Notes

There are two types of smoke detectors. One type is connected to a building's fire alarm system and acts only as a smoke sensor; sending a signal to the fire alarm panel which sounds an alarm throughout the building. The second type is the "*single station battery powered smoke detector.*" These are not connected to a building's fire alarm system and are similar to those found in private homes. The single station type has the sensor, and alarm built into one unit. These units are often hard wired to the electrical system but have a battery to allow the detectors to operate during a power outage.

In apartments and hotels, it is common to have detectors in hallways that are part of a building-wide alarm system and single station detectors in guest rooms. This prevents nuisance alarms caused by cigarette smoke from evacuating an entire building.

Single station battery powered smoke detectors typically have a design life of 10 years before they should be replaced. Some manufacturer's may have different replacement intervals.

See also Carbon Monoxide Detectors.

SPRINKLERS

See: Lawn Irrigation, Fire Sprinkler Systems.

SUMP PUMPS

Sump pumps remove ground or rain water from basements, crawlspaces, or other areas where standing water is a problem. Most sump pumps discharge water onto the ground away from the building while some discharge into a building's storm water system.

In some locations, sump pumps can be considered critical equipment and a sump pump installation can include backup pumps, automatic lead/lag controls, and failure alarms.

PM Tasks

To check the operation of the sump pump you can usually push down on the float or submerge the switch to turn the pump on. Verify that the water level drops in the sump pit. Failing check valves are a common problem with sump pumps. Water should not flow back into the sump pit once the pump turns off. If lead/lag pumps are present, make sure each pump operates and submerge any alarm switches to test operation of the alarm.

Recommended PM Frequency

Monthly: Inspect and test.

SWIMMING POOLS

In all except the most southern states, outdoor swimming pools are operated during only part of the year. Indoor swimming pools can be operated year round in all climates.

PM Tasks

When shutting down the pool for the winter, make sure to check the following items: remove the drain plugs from pumps and filters, to prevent damage from freezing. The water level in the pool should be pumped down to several inches below tile coping to prevent rising ice from damaging the coping at the top of the pool wall. Lowering the water level in the pool also prevents water from entering the scuppers where it could freeze and cause damage. Lines to scuppers and wall returns should be blown out with compressed air toward the pool and expansion plugs should be installed to prevent water from entering these lines during the winter. Swimming pool antifreeze can be used instead of draining and blowing out lines. Add algaecide and place a pool cover over the pool to prevent leaves, and trash from blowing into the pool and to make the pool less dangerous. A small pump made for the purpose should be placed in the lowest part of the pool cover and operated as needed on warm winter days to remove water from the top of the cover. Do not leave an in ground swimming pool completely empty over the winter. Heavy rains can saturate the ground causing a swimming pool to float like a large boat.

When opening a pool for the season, first drain the pool completely to allow an inspection to be performed. Inspect the pool for cracks or damaged paint on the inside of the pool. Many facilities scrape and re-paint the inside of concrete pools with rubber waterproofing paint once per year. Pools in good condition and fiberglass pools do not need this treatment.

Every day during the operating season, pools need to be cleaned and inspected for hazards such as loose tile, loose diving platforms, non-latching fence gates, operation of pool alarms, exposed electrical wiring, or damage to any pool equipment. Check pool lights. Water and electricity can be a dangerous combination and underwater lights are sealed to prevent water from entering the light fixture. If you see a waterline behind the lens, close the pool immediately and replace the gasket. Make sure main drain safety covers are in place to prevent anyone getting trapped underwater by suction entrapment. Pool Life-guards can usually perform these inspections as part of their duties freeing up skilled maintenance staff to do other jobs. Outside profes-sionals are often used to open and close pools for the season with in-house Certified Pool Operators (CPOs) handling daily operations including chlorine and pH testing, adding chemicals, and cleaning filters.

Many jurisdictions require 5-year electrical bonding inspections to be performed by a licensed electrical contractor.

Recommended PM Frequency

Daily: Visual safety inspection (during season).

Annually: Season start up in the spring or summer.

Annually: Season shut down in the fall.

Every 5 years: Electrical bonding inspection (depending on local codes).

Technical Notes

In many states, public swimming pools can only be operated by a CPO. CPOs must complete three days of training in water testing, chemical treatment, pool safety, and other issues related to the safe and sanitary operation of swimming pools.

TERMITE INSPECTIONS ☎

In all except for the northern-most parts of the U.S., subterranean termites are a threat to wooden structures. In the southern-most regions of the U.S., the Formosan termite, an extremely aggressive wood destroyer, does millions of dollars of structural damage every year.

In areas susceptible to termite damage, termite inspections should be performed so that treatment can be performed before significant damage occurs. In most areas termite inspections are performed only by licensed inspectors or by licensed pesticide applicators.

PM Tasks

The frequency of termite inspections depends on your area's termite damage risk. Northern states may not need any termite inspections while areas with Formosan termite infestations may need annual inspections.

Recommended PM Frequency

Every 5 years: Most parts of the US with risk of termite damage

Every 3 years: Southern US without local infestations of Formosan termites

Annually: Areas of US with known Formosan termite infestations.

Technical Notes

There are a few treatment options available for termites. The traditional method is to create a chemical barrier around the perimeter of the structure by saturating the soil with a termite killing insecticide. During new construction, backfill material is often drenched with a barrier chemical prior to backfilling. In existing construction, a method called "rodding" is often used where a long tube is used to inject the chemical into the soil around the building perimeter. This rod is forced into the soil approximately every 12 inches along the foundation. The goal is to create a continuous chemical barrier. Drench treatments last approximately 5 years.

Since no chemical barrier is ever completely continuous, termite inspections should continue even after treatment. If active termite infestation is found after treatment, retreatment may be required.

In recent years, termite bait systems have become common. Bait systems are often used as a preventive measure as well as a treatment for existing infestations. With the bait system, bait stations are placed in the ground around the structure. These bait stations contain wood or another form of cellulose that termites will eat. These bait stations are inspected for termite activity and when activity is found, the cellulose bait is replaced with a chemical insecticide which the termites take with them back to the colony.

TRANSFER SWITCHES

See Emergency Generators.

TRUCKS

See Vehicles.

VACUUM CLEANERS (& OTHER HOUSEKEEPING EQUIPMENT)

When compared to chiller plants, roofs, or fire sprinkler systems; maintaining vacuum cleaners and other housekeeping equipment may seem trivial. However, poorly maintained floor care equipment results in dirty floors. Dirt on hard flooring such as VCT causes wear and makes floors look dingy. Dirt in carpet fibers causes premature wear of carpet. Keeping housekeeping equipment in good condition helps keep the rest of the building in good condition.

Equipment such as floor buffers, vacuum cleaners, and carpet extractors get hard use. When performing PM on housekeeping equipment, it can be surprising how many have broken beater bar belts, clogged filters or hoses, full vacuum bags, frayed cords, broken switches, or other problems. These machines are still being used daily even though they're no longer doing their job well.

PM Tasks

Make sure all parts of the equipment are working properly. Check any vacuum paths for obstructions, change motor brushes and oil motors when appropriate. Be prepared with replacement power cords since this is one of the most common repairs needed during PM. Make sure any safety switches are operating (floor buffers should require both handles to be pressed to operate.)

Recommended PM Frequency

Monthly: Inspect all housekeeping equipment.

Technical Notes

Each piece of housekeeping equipment should be given a unique number and repairs recorded. If the same machine keeps showing up in your shop with the same problem, a repair record will show you that you need to look deeper for the cause.

VALVES

Water valves can become stuck from rust and corrosion if the valves sit for a long time without being operated. Many valves sit unused for years and may not function when needed in an emergency. In an ideal world, it would be nice to operate every valve in your building twice a year so that no valve becomes corroded in the open position. However, in most buildings that would be a daunting task and the locations of all valves aren't always known.

The main valves for a building and any branch valves serving large areas of the building should be operated regularly since these are the valves that will be needed in an emergency. This rule applies to valves for domestic water, heating water, and natural gas.

PM Tasks

Lubricate stem with light oil. Fully open and close valve. Tighten

Figure 11-18. Brass tags are one way that valves are numbered for future identification.

packing nut or replace packing if there is any dripping around the valve stem. Do not oil or loosen packing nuts on gas valves.

Recommended PM Frequency

Semi-annually.

Technical Notes

Each valve in your building should have a unique valve number attached to it with a piece of wire or ball chain. A "valve chart" listing each valve, its location, and its function will be very helpful when trying to shut off water in an emergency. I recommend creating such a valve chart and adding a valve number tag to each valve when doing your building inventory. Without a valve chart, it's unlikely that your PM tech will find all of the valves in your building. A set of plumbing system blueprints will help you in creating a valve chart.

Many valves also have the valve's function and "normally open" or "normally closed" printed on the valve's tag. Boiler rooms are often crowded with pipes for hot water, domestic water, cooling tower loops, heat exchangers, heating water, makeup water, fire suppression, booster pumps, and others. Having each valve's function printed on the valve's tag can make troubleshooting and operating each of these systems much easier.

VEHICLES

It wasn't long ago that cars and trucks all needed to have the oil, oil filter, and air filter changed, and chassis greased every 3000 miles, tires rotated every 10,000 miles, and transmission fluid, filter and engine coolant changed every 50,000 miles. In the last two decades a lot has changed in the service intervals required for vehicles.

Many vehicles require oil changes every 5000, 7000, or even 10,000 miles now. Many vehicles do not have a scheduled service interval and instead monitor engine temperature, number of engine revolutions, and driving conditions to determine when it's time to perform service.

Since vehicle PM is based on miles driven or hours of operation, instead of calendar date, it is a difficult item to schedule accurately on a PM calendar. If you have your vehicles serviced at an auto shop instead of in-house, they will usually affix a sticker to the window reminding the driver of the next required service. If you perform service in-house you can purchase these stickers from many auto parts stores or use any type of sticker or hang tag to indicate the next required service.

PM Tasks

The service requirements for vehicles vary greatly from vehicle to vehicle. Consult the vehicle's owner's manual for the specific requirements for your vehicles.

The only item that can be entered accurately on your PM schedule is dates for renewing registrations, or having vehicles inspected for safety/emissions according to state laws.

Recommended PM Frequency

See manufacturer's literature.

VENTILATION FANS

See Exhaust Fans.

VINYL COMPOSITION TILE (VCT)

See Resilient Flooring Care.

WASHERS (COMMERCIAL LAUNDRY)

See Laundry.

WATER SOFTENERS

Water softeners are often used in buildings that have "hard water." Hard water contains minerals which can crystallize inside plumbing and which can cause problems with the function of soaps and detergents.

PM Tasks

Water softeners need to have water softener salt added to the salt tank. At the same time, any filters for iron or other minerals that are part of the water treatment system, these should be checked or replaced.

Recommended PM Frequency

Monthly: Add salt to system. Inspect system for any leaks, check timer or water meter operation, check condition of filters.

Water filter replacement interval will need to be determined by volume of water used, concentration of iron or other minerals present, and the size of the filter used.

Technical Notes

When minerals such as calcium, magnesium carbonate, and manganese are present in high concentrations in water that water is considered to be "hard." Water with more than 3.5 grains per gallon (GPG) of minerals is considered to be hard water. Water with more than 10.5 GPG is considered to be very hard.

Dissolved minerals can re-crystallize creating scale that blocks plumbing or leaving deposits on sinks, toilets, and clogging shower heads. Scale on the inside of water heaters, boilers, and heat exchangers acts as an insulator slowing heat transfer and reducing the efficiency of heating and cooling equipment. Minerals in hard water also react chemically with soap preventing soap from lathering and creating a gummy deposit called scum. Hard water also requires much more soap to be used to do the same job.

Water softeners remove these dissolved minerals by replacing them with dissolved sodium from sodium chloride (salt). A water softener's water treatment tank is filled with zeolite beads. Zeolite materials can hold charged molecular ions on their surface. By soaking these beads in salt the beads are made to be covered with sodium ions. As water passes over these beads, the calcium or magnesium ions in the water are replaced with sodium ions and the Ca and Mg ions are left on the surface of the beads. When all of the sodium ions are depleted and the beads are covered with Mg or Ca ions, the zeolite beads must be soaked in salt brine again to recharge them with sodium ions. This recharge cycle is handled automatically by a timer or water meter.

Appendix

SAMPLE PREVENTIVE MAINTENANCE RECORD FORMS
- Emergency Generator Testing Log
- Room PM Checklist
- Smoke Detector Testing Log

TROUBLESHOOTER'S CREED

TRUISMS

Emergency Generator Test Log

Testing Completed By: ____________________ Year________________

	Jan	Feb	Mar	Apr	May	Jun	Jul	Aug	Sep	Oct	Nov	Dec
Test Date												
Pre-Start												
Battery water level												
Crankcase Oil												
Fuel Oil Level												
Coolant Level												
Engine Running												
Oil Pressure												
Oil Temperature												
Battery Volts												
Water Temperature												
Generator Running												
AC Volts Φ1												
AC Volts Φ2												
AC Volts Φ3												
AC Amps Φ1												
AC Amps Φ2												
AC Amps Φ3												
RPM												
Frequency												
Full Load Test (Y/N)												

Hotel Guest Room Preventive Maintenance Checklist

Room ______________________

Date ______________________

Completed By ______________________

P = Paint R = Repair

N = Needs Replacement OK = OK

	Date	Date	Date	Date
Bathroom	Qtr1	Qtr2	Qtr3	Qtr4
Bath mat				
Cault/Grout				
Ceiling				
Commode angle stops				
Commode bolt caps				
Comode caulking				
Commode lever				
Commode lid/seat				
Commode supply line				
Commode tank				
Curtain rod/door/curtain				
Door frame				
Door hinges				
Door knobs/strikeplate				
door latch				
Door robe hook				
Door stop				
Door threshold				
Front Door				
Room number				
Frame				
Threshold				
Lock lever				
Lock strike plate				
Lock latch				
Dead bolt lever				
Security night latch				
Door viewer				
Door stop				
Smoke strip				
Closer				
Emergency Evac. Plan				

	Date	Date	Date	Date
Bathroom (cont.)	Qtr1	Qtr2	Qtr3	Qtr4
Drain Cover/lever				
Exhaust vent/fan				
Faucets/controls				
Floor, wall tile, marble				
Lights				
Mirror				
Outlet, GFI, coverplates				
Shower head/plate				
Sink, faucet, stopper				
Soap Dish				
Switches, cover plates				
Towel rack				
Tub				
Wall vinyl				
Air				
Thermostat				
Filter				
Filter vent cover				
Coils				
Drain pan				
Fan				
Compressor				
Supply lines				

continued on next page

Hotel Guest Room PM Checklist Page 2 of 2

	Date	Date	Date	Date
Bedroom	Qtr1	Qtr2	Qtr3	Qtr4
Smoke detector				
Balcony rail, floor				
Sliding door , window				
Sliding lock				
Sliding stop				
Sliding frame				
Drapes, rod, hooks				
Closet door, track,				
Ceiling				
Wall vinyl				
A/C vent cover				
Headboards				
End tables				
Amoire doors				
Amoire drawers				
TV, cable box				
TV remote control				
Lamps, shades, bulbs				
Switches, cover plates				
Outlets, cover plates				
Phone				
Vanity mirror, lights				
Sink, faucet, stopper				
Towel bar				
Bed, frame, box spring				
flip Mattress				
Clock radio				
Pictures, art work				
Carpet				
Hall, Bedroom Door				
Hinges				
Lever handle				
Latch				
Strike plate				
Door stop				
Electronic lock battery				

	Date	Date	Date	Date
Living Room	Qtr1	Qtr2	Qtr3	Qtr4
Smoke detector				
Walls				
Ceiling				
Carpet				
Mirror				
Pictures, artwork				
Ceiling vent				
Lamps, shades, bulbs				
Switches, cover plates				
Outlets, cover plates				
Table, chairs				
Sofa bed, mechanism				
Lounge chair				
End/coffee tables				
Phone				
TV, cable box				
TV remote control				
Drapes, rods, hooks				
Adjoining door, latch				
Window				
Kitchen, Wetbar				
Microwave				
Cabinets, doors				
Walls				
Ceiling				
Paper towel holder				
Lights				
Carpet				
Refrigerator				
Outlet, cover plates				
Switches, cover plates				
Bar sink				
Bar sink counter				
Bar sink stopper				
Bar sink faucet				
Bar sink strainer				

Monthly Smoke Detector Testing Log

Date ______________________ Completed By ______________________

Room #		Room #		Room #		Room #	
	□ okay □ replace □ change battery		□ okay □ replace □ change battery		□ okay □ replace □ change battery		□ okay □ replace □ change battery
	□ okay □ replace □ change battery		□ okay □ replace □ change battery		□ okay □ replace □ change battery		□ okay □ replace □ change battery
	□ okay □ replace □ change battery		□ okay □ replace □ change battery		□ okay □ replace □ change battery		□ okay □ replace □ change battery
	□ okay □ replace □ change battery		□ okay □ replace □ change battery		□ okay □ replace □ change battery		□ okay □ replace □ change battery
	□ okay □ replace □ change battery		□ okay □ replace □ change battery		□ okay □ replace □ change battery		□ okay □ replace □ change battery
	□ okay □ replace □ change battery		□ okay □ replace □ change battery		□ okay □ replace □ change battery		□ okay □ replace □ change battery
	□ okay □ replace □ change battery		□ okay □ replace □ change battery		□ okay □ replace □ change battery		□ okay □ replace □ change battery
	□ okay □ replace □ change battery		□ okay □ replace □ change battery		□ okay □ replace □ change battery		□ okay □ replace □ change battery
	□ okay □ replace □ change battery		□ okay □ replace □ change battery		□ okay □ replace □ change battery		□ okay □ replace □ change battery
	□ okay □ replace □ change battery		□ okay □ replace □ change battery		□ okay □ replace □ change battery		□ okay □ replace □ change battery
	□ okay □ replace □ change battery		□ okay □ replace □ change battery		□ okay □ replace □ change battery		□ okay □ replace □ change battery
	□ okay □ replace □ change battery		□ okay □ replace □ change battery		□ okay □ replace □ change battery		□ okay □ replace □ change battery
	□ okay □ replace □ change battery		□ okay □ replace □ change battery		□ okay □ replace □ change battery		□ okay □ replace □ change battery
	□ okay □ replace □ change battery		□ okay □ replace □ change battery		□ okay □ replace □ change battery		□ okay □ replace □ change battery
	□ okay □ replace □ change battery		□ okay □ replace □ change battery		□ okay □ replace □ change battery		□ okay □ replace □ change battery

THE TROUBLESHOOTER'S CREED

Use the following rules to help in diagnosing maintenance problems.

Understand the Sequence of Operation—If you don't know first how a system should be working you will have a very hard time figuring out why it isn't.

Start with the Most Obvious—The most obvious cause is usually the right one. Don't start looking for complex causes until the most simple ones have been eliminated.

Fix the Problem You Know About—If you know part A is bad but it seems like part B may also be acting up, replace part A first and test.

Assume Nothing, Verify Everything—Someone's description of the problem can be a help but is often wrong. Check all the information you have been given for yourself.

Rely on All of Your Senses—You can often hear, smell, or feel something that you would have otherwise missed.

Divide the System in Half—Large or complex mechanical and electrical systems can usually be separated into smaller systems and each of these sub-systems operated independently. You can then divide the non-working sub-system again and again until the cause of the problem is isolated.

TRUISMS

ONE

For PM to be successful, you must have an attitude of continual improvement

TWO

A dollar saved through PM is as good as a dollar earned through any other business activity.

THREE

You cannot save money by skimping on PM

FOUR

The only right way to perform PM is do follow the equipment manufacturer's maintenance procedures.

FIVE

You are not behind, don't try to catch up

SIX

If you fail to remind management of your PM program and its baby steps toward success, it will be quickly forgotten and other priorities will take over.

SEVEN

Training someone and having them leave is better than not training them and having them stay

EIGHT

People do what you inspect, not what you expect.

NINE

Good is the evil enemy of great

GLOSSARY OF PREVENTIVE MAINTENANCE TERMINOLOGY

alkyd—a synthetic binder used in oil base paint. Also another name for oil-based paint.

asbestos—a mineral fiber used as insulation, fire proofing, reinforcing fiber, and other applications in construction materials. Asbestos fibers are now known to cause asbestosis and cancer.

asset—any owned property. Specifically any equipment or building component.

BAC—see building automation control.

bearing—the supporting guide for a rotating or sliding machinery component.

best maintenance practice—a commonly accepted industry maintenance practice that is used when the manufacturer's recommendations are not available.

binder—the component of paint that polymerizes and develops into the final film of paint. Sometimes called the "solids" of the paint.

biological contaminants—pollutants from living things such as pet dander, rodent droppings, mold, bacteria, etc.

bitumen—asphalt or tar-based sealant and adhesive used between layers of roofing felt in many commercial roofs. Can be applied hot or mixed with solvents to make the bitumen liquid enough to mop onto the roofing layers.

BMP—see best maintenance practice.

Btu—British thermal unit. A measure of thermal energy. 1 Btu is the amount of heat required to raise one pound of water one degree Fahrenheit.

building automation control—A heating and cooling control system where temperature, humidity and occupancy sensors send data to a central computer which directly controls all heating and cooling equipment in a building.

building envelope—the outside surfaces of a building. Walls roofs windows doors etc.

built-up roof—a commercial roofing system consisting of several layers (usually 3 or 4) of roofing felt embedded in layers of asphalt or tar. (See also modified bitumen roof.)

BUR—see built-up roof

call back maintenance—maintenance work that needs to be done again often because it wasn't done right the first time.

CLAIR—clean, lubricate, adjust, inspect, repair. An acronym to help PM mechanics remember the proper procedure for performing PM.

CMMS—computerized maintenance management software.

cold galvanizing compound—zinc-based paint formulated to protect metal from rusting.

corrective maintenance—repair. Fixing something after it has broken.

DDC—see direct digital control.

deferred maintenance—maintenance work that is not being done at this time.

direct digital control—a heating and cooling control system where temperature, humidity, and occupancy sensors send data to a central computer which directly controls all heating and cooling equipment in a building.

direct expansion—a type of air conditioning equipment where the refrigerant exchanges heat directly with air via a cooling coil, e.g. "DX coil."

downtime—time that equipment is not working due to some equipment failure.

dust spot efficiency rating—method of testing and reporting how well air conditioning filters remove particles from the air.

dust weight arrestance rating—method of testing and reporting how well air conditioning filters remove particles from the air.

DX—see direct expansion.

EAC—see equivalent annual cost.

elastomeric wall coating—an architectural coating that forms a very thick flexible film.

emergency maintenance—maintenance work that needs to be done immediately.

envelope—see building envelope.

EPDM—ethylene propylene diene monomer. The type of rubber used in the majority of single-ply commercial roofs. Often referred to as a rubber roof.

equipment data sheet—a form used to record important information about each piece of equipment in a facility.

equivalent annual cost—used to compare the costs of repair vs. replacement options. Used as a tool in "life cycle costing."

EWC—see elastomeric wall coating.

flashing—roofing component that provides a water shedding transition between two different surfaces.

flat roof—a roof having a pitch of less than 1/4" rise for every 12" horizontally.

grease—lubricating oil mixed with a thickener. Used to lubricate slow moving equipment or rolling element bearings.

group re-lamping—replacing an entire section of fluorescent or HID lamps at one time as they reach the end of their expected life.

HID—see high intensity discharge lighting.

high intensity discharge lighting—type of lighting that is more efficient than traditional incandescent lights. Typically used for exterior or high bay lighting applications. Includes metal halide, mercury vapor, and high pressure sodium lighting.

HVAC—the trade involved in heating ventilation and air conditioning.

HVAC/R—heating ventilation air conditioning and refrigeration.

hydrodynamic pressure—fluid pressure generated from the rapid compression of a fluid. In fast rotating bearings hydrodynamic oil pressure keeps bearing surfaces from touching, eliminating surface to surface wear.

hydrostatic pressure—fluid pressure generated by a source outside of the fluid. In slow rotating or heavily loaded bearings, oil pumps may be used to create hydrostatic pressure.

IAQ—see indoor air quality.

indoor air quality—the presence or absence of airborne pollutants inside a building.

infant mortality—the concept that brand new equipment or parts are more likely to fail than those parts that have been in service for some time due to manufacturing defects.

infrared imaging—see thermal imaging.

inverted roof—a roofing system where insulation board is applied on top of the roofing membrane to protect the roof membrane from physical damage and temperature extremes.

JCAHO—Joint Commission on the Accreditation of Healthcare Organizations, organization that sets voluntary standards for healthcare centers that exceed state and federal standards.

latex paint—paint that uses water soluble plastic or polymer binders and water as the carrier.

lead—a toxic heavy metal once used extensively in architectural paints.

life cycle costing—comparing the cost of various maintenance repair or replacement options over the life of the various options. Requires converting all options to one common unit of time.

low pitch roof—a roof having a pitch of less than 2″ of rise for every 12″ horizontally but more than 1/4″ rise for every 12″ horizontally.

lubrication—the application of grease or oil to a machine to reduce friction.

MBR—see modified bitumen roof.

MERV—see minimum efficiency reporting value.

minimum efficiency reporting value—method of testing and reporting how well air conditioning filters remove particles from the air.

modified bitumen roof—commercial roofing system consisting of several layers (often 4) of roofing felt that is factory impregnated with asphalt or tar bitumens. Modified bitumen roofs get their name from additives (modifiers) that are added to the tar or asphalt to improve roof performance. See also built-up roof.

mold—microscopic fungi that feed and grow on organic material. Often associated with poor indoor air quality within buildings.

neutral axis—the imaginary line through a drive belt where no compression or expansion of the material occurs when the belt flexes. Usually located where the reinforcing cords are located.

oil analysis—laboratory analysis of used lubricating oil to determine the condition of the equipment the oil sample was taken from. Often used as part of a predictive maintenance program.

oil-based paint—paint that uses petroleum-based solvents, usually mineral spirits as the paint carrier or solvent.

orifice—type of refrigerant expansion device consisting of a disk with a small hole to allow a metered amount of refrigerant to pass.

paint—a thin film coating of polymer applied as a liquid.

pathway—the route of travel of a pollutant. Specifically related to indoor air quality.

PDA—personal data assistant. A small hand held computer.

PdM—predictive maintenance. Maintenance that is performed when indicated by some measurable wear factor.

pesticides—chemicals used to kill pests.

pigment—the component of paint that provides the color or rust inhibitive properties to paint.

pitch—a measurement of the slope of a roof given in inches of rise for every 12″ horizontally.

pitch pocket—a roofing component that is filled with asphalt or tar pitch to seal around pipes conduit or other penetrations.

pollutant—an airborne contaminant. Related specifically to indoor air quality.

ponding—standing water on a flat or nearly flat roof.

predictive maintenance—maintenance that is performed when indicated by some measurable wear factor.

preventive maintenance—a scheduled program of regular inspections adjustments lubrication or replacement of worn or failing parts in order to maintain an asset's function and efficiency.

primer—a type of paint engineered to work as an undercoat to improve adhesion and appearance of a top coat.

PTAC—package terminal air conditioner. Small self contained air conditioner units that are installed in a sleeve through a wall. Often found in hotel rooms.

pulley—wheel that drives or is driven by a drive belt. See also sheave.

pulley misalignment—when two pulleys are parallel but not in the same plane. Causes excessive heating and premature wear of drive belts.

purlin—a roof framing member applied on top of and perpendicular to the rafters.

radon—a colorless odorless radioactive gas that results form radioactive decay of uranium in rock. Radon can infiltrate buildings where it can be a cause of cancer.

reactive maintenance—see corrective maintenance.

reliability engineering—a field of engineering focusing on predictive maintenance and machinery reliability.

retro-commissioning—having a building and particularly the HVAC systems tested, inspected, and adjusted to maintain the energy savings designed into the system when it was new.

return on investment—financial term for the ratio of money invested to money earned from the investment given as a percentage.

ROI—see return on investment.

RPM—rotations per minute.

RTU—rooftop air conditioning unit.

run to failure—operating a piece of equipment with no intention of doing any maintenance until the equipment breaks down. Often used for inexpensive, easily replaced equipment.

rust convertor—a protective coating including tannic acid which converts rust to a hard protective stable compound.

shaft misalignment—when two pulleys are not parallel or in the same plane. Causes excessive heating and premature wear of drive belts.

sheave—wheel that drives or is driven by a drive belt. See also pulley.

sick building syndrome—phenomenon where building occupants suffer health symptoms while in a building that leave when the person is not in the building.

solvent—the component of paint that evaporates away leaving a film made of the binder and pigments.

thermal imaging—a non destructive method of testing equipment using thermal imaging cameras that can see and measure temperature. Often used as a predictive maintenance tool.

thermoplastic—a plastic material that can be softened by heat. Several of these plastics are used as single-ply membrane roofing materials. A common example would be PVC (polyvinyl chloride).

thermostatic expansion device—a type of refrigerant expansion device. Uses a temperature sensing bulb to adjust the size of the expansion orifice to maintain a constant refrigerant temperature at the evaporator coil.

ton—a measure of cooling capacity. A ton of cooling is the amount of cooling required to produce one ton of ice in a 24-hour period. The same as 12,000 Btu.

TXV—see Thermostatic Expansion Device.

ultrasonic testing—a non destructive method of inspecting machinery components, such as motor bearings, by using a listening device to hear sounds normally out of the range of human hearing.

urethane—a group of architectural coatings based on reacting urethane with a catalyst to harden.

vibration analysis—a non destructive method of testing machinery components such as motor bearings by measuring and analyzing component vibrations.

viscosity—a measurement of a fluids resistance to flow.

VOC—see volatile organic compound.

volatile organic compound—organic materials that vaporize readily at normal temperatures. Often a component of paints, solvents and other maintenance products. VOCs can contribute to respiratory or environmental problems.

work order—a written or computerized request for maintenance work to be completed.

Index

Symbols
80-20 rule 42

A
abatement 166
access doors 35
access to equipment 18
acrylic 177, 178
adhesion 182
AEDs 209
aesthetics 39
air compressors 207
air conditioning 125, 208
air contaminants 135
air filter 135
air handler 130, 131
algae control tablets 137, 227
alkyds 176, 177
allergens 135
allergic 167
    reactions 160, 161, 188
allergist 167, 170
alligatoring 192
aluminum 185
    soap 94
angular velocity 86
Ansul® 253
    systems 209
anti-scald 250
APP 107
appearance 40
application temperature 191
asbestos 160
asbestosis 171
ASHRAE 135, 163
asphalt 105, 106
asset 5
asthma 160, 161, 167
atactic propylene 107
atmospheric vacuum breaker 213
attitudes 69
attitude of continual improvement 17, 70
automated controls 144
automated external defibrillators 209
automatic air eliminator 210
automatic air vents 210
AVB 213

B
backflow prevention devices 212, 255
backlog 26, 28
ball roller bearing 96
barium soap 94
base flashing 114
bathtub chart 11
bead board 112
bearing 83
    clearance 87
    journal 84
belt length 152
benchmarks 23
best maintenance practices (BMP) 50
binder 176
bitumen 105
bleach 193
bleed through 182, 192

blistering 192
blowers 137
blueprints 40
boilers 215
bonding primers 184
boron nitride 92, 93
breakdowns 8
British thermal units (Btus) 126
budgeting 22, 27
buffing 266
building automation controls (BAC) 72, 145
built-up roofing (BUR) 104, 105
burners 138
burner efficiency 138
burn in 11

C

calcium 281
    soap 93
calendar 57
call-back maintenance 16
capital equipment 40
capital improvement 28
capital project 28
cap sheet 106
carbon dioxide (CO_2) 164
carbon monoxide 138, 173
    detectors 217
carcinogens 188
carpet care 218
carpet extractors 278
carrier 176
cars 280
caulk 186, 220
cellular glass board 112
cell phone 56, 59
centipoise 89
centistokes 89
centrifugal compressor 143
centrifugal fans 137
certified 76
    playground safety inspector 262
certified pool operators (CPOs) 275
chalking 179, 192
chemical treatment 142
chillers 126, 142, 223
chipping (flaking) 193
chromate 181
circulator pumps 224
CLAIR 50
classic cross section belts 150
closed circuit 127
CMMS 21, 44, 48, 51, 53, 58
CO 173, 217
CO_2 170
coal tar 106
    pitch 105, 106
coatings 175
codes 42
coefficient of performance 130
cold galvanizing compounds 181
combustion gasses 138
combustion heating 130
comfort issues 163
commissioning 72
communication 172
complaints 14
comprehensive maintenance plan 27
compressor 127, 130, 131
computer 48, 53, 58
condensate 136
    drain 134, 226
    equipment 136
    loop 225

pump 137
water loop 139
condenser 130
coil 128, 130, 133
fan 130
condition-based 9
consultant 75
containment barrier 168
contaminants 83, 94, 99, 127, 160
continual improvement 36
contractor 75
controls narrative 145
cooling fins 133
cooling loops 139
cooling towers 139, 142, 225
COP 130
coping 119
core samples 10, 121
corrective maintenance 13
corrosion 142, 180
costs 5, 21, 22, 34
counter flashing 115
creeping 193
cyberspace 54
cylindrical roller bearing 96

D

damaged bearing surfaces 94
damaged seals 100
data 169
deferred maintenance 15, 28
deflection 105, 111
depreciation 33
design 17
detention ponds 267
developmental problems 189
direct digital control (DDC) 72, 145
direct expansion (DX) 128
discharge 127
dishwashers 228
disposal 189
costs 33
distribution panels 230
doors 229
double check valve 213
down time 6, 9, 33
drain plugs 97
drive belts 149
dryers 254
dry cooling tower 139
dust 135, 160
dust spot efficiency rating 135
dust weight arrestance rating 135
DX unit 130

E

EAC 30
economizers 131
efficiency 59
efflorescence 193
elastomeric wall coatings 179
electrical bonding inspections 275
electrical problems 10
electrical systems 230
electric heat strips 130
electronic expansion valves 128
elevators 232
emergency generator 233
test log 284
emergency lighting 235, 242
emergency maintenance 16
enamels 178
energy costs 7
energy crisis 159
energy efficiency 33
engineering economics 32
envelope 10

environment 125
environmental consultant 171
enzymes 248
EPA 125
EPDM 107, 108
    rubber membranes 104
epoxy paints 180
epoxy sealers 137
equipment data sheet 42, 50
equipment history 29
equivalent annual cost 30
escalators 232
evaporative condenser 139
evaporator blower 130
evaporator coil 128, 130, 134
EWCs 179
exhaust fans 173, 236, 251
exhaust hoods 251
exit hardware 242
exit signs 235, 242
expansion device 128
expansion valve 130
expectations 39
extension cords 231, 242
extreme pressure greases 92, 93
EXVs 128

F

fasteners 113, 114
fertilizer 255
fiberglass batt insulation 113
filter 131, 134, 173
    efficiency 135
fin comb 133
fire alarm systems 236
fire extinguishers 237
fire hydrants 239
fire inspections 240
fire protection systems 243
flag poles 244
flashing 113, 114
flat roofs 104
flood coat 106
floor buffers 278
floor drains 172, 245
floor finish 266
FOG interceptors 247
folklore 79
fractional horsepower belts 150
freezers 265
frequency 48, 51
fresh air 164
friction 83
funding 21
furnace 246
    filters 135

G

gale strength 117
galvanized iron 185
gas pressure 138
Gates belts 149
glass fiberboard 113
global warming 125
goals 39
government regulations 42
graphite 92, 93, 99, 259
gravel stops 120
grease 92, 95
    fitting 96
    gun 96
    interceptors 247
    life 96
    plugs 97
    port extension 35
        kits 18, 100
    pressure release fittings 100
    traps 247

greenhouse gasses 188
grounds care equipment 257
ground fault circuit interrupters (GFCI) 231
group re-lamping 258, 259

H
hazardous waste 178, 190
headaches 188
hearing loss 189
heating equipment 138
heat exchanger 138, 142
heat pumps 129, 141
heat strip 138
HEPA 168
hermetic 131
hermetically sealed compressor 131
hermetic compressors 132
HID 11, 258
high temperature grease 93, 94
holidays 193
hot water 248
hour meter 52, 53
human resource (HR) 68
humidity 138, 163, 170
  removal 136
HVAC 75, 125
  equipment 42
  systems 125
hybrid cooling tower 139
hydrodynamic 85
  pressure 84, 85

I
ice machines 250
ID number 45, 56
in-house 38
indoor air quality (IAQ) 135, 159, 188
industrial belts 150
industrial hygienists 171
infant mortality 11
inflation 33
inner race 99
inspecting paint 194
inspection 53
installation errors 11
insulation 110
  board 104
intensity discharge 258
internal combustion engines 94
internet 54
inventory 40, 44
inverted roof 110
investment 21
iron tannate 187
ISO number 89

J
jacks of all trades 67, 75

K
kidney 188
  damage 189

L
labor hours 54
laptop computer 56, 59
latex-based paint 176
lawnmowers and grounds care equipment 257
lawn irrigation 255
lead 181
  abatement contractors 189
  paint 186, 189
Legionella 171
library 49

licensed 76
lighting 258
lime soap 93
linoleum 266
lithium soap 94
liver 188
location 44, 47
lockers 260
locks 259
long-range facilities plan 27
loss of gloss 193
low slope roof 104
low water cut off 216
lubrication 83
  theory 83
lumen maintenance 11, 258

M

magnesium carbonate 281
makeup water 142
management 23, 61
manganese 281
manufacturer 49
manufacturing defects 11
masonry 184
matched belts 154
material costs 35
MBR 106
MERV 135
  ratings 165
metals 184
mildew 186, 193
mildew-resistant paints 181
mildewed 220
mineral board 112
mineral oil 91, 94
mineral spirits 176
minimum efficiency reporting value 135
mistrust 169, 172
mixing valves 250
modified bitumen 106
  roofing (MBR) 104
mold 111, 161, 163, 167
  spores 135
molybdenum disulphide 92, 93, 99
mopping 266
motivation 69
muriatic acid (hydrochloric acid) 193
mycotoxins 167
myths 79

N

narrow profile belts 150
nausea 188
needle roller bearings 96
nervous system damage 188, 189
neutral axis 152
NFPA 101 229
notched belts 153
no dollar limit warranties 116
nurse call system 260

O

oil 83, 84
  analysis 9, 95
  based paint 176
  burners 138
  film 86
  pressure 84, 86
  reservoirs 94
  sight glass 132, 133
Oilite bearings 98
orifice 128
Osborn Reynolds 83
OSHA 164

outer race 99
outside air dampers 131
outside contractors 6
outsourcing 75
overtime 21, 35
ozone 188
    layer 125

P

package system 130
paint 175
panic 172
    hardware 229
parking lots 261
parts 54
    inventory 53
pathways 160, 161, 165
PDA 56, 59
people 67, 160, 162, 166
permissible exposure limit 164
pesticides 161, 255
physical barrier 165
pitch pockets 121, 122
plain bearings 98
planning 17
playground equipment 262
PM cart 270
pneumatic 207
pollen 135
pollutant 160, 165
    source 165
polyepoxides 180
polyisocyanurate 112
polymer 177
polystyrene 112
polyurethane 112, 176, 180
ponding 105, 110, 111
power strips 242
predictive maintenance 9
pressure drop 136
pressure relationships 165
pressure release plugs 100
pressure vacuum breaker 214
preventive maintenance 1
primers 182, 183
profits 22
prorate 117
prorated 116
public places 40
PVB 214
PVC (polyvinyl chloride) 108

R

radon 160
    mitigation equipment 263
railroad 84
rain gutters 264
re-greasing interval 96
re-lamping 10
reactive maintenance 13
rebate programs 33
refrigerant 125, 127
    oils 132
refrigerators 265
regulations 42
regulatory compliance 64
reliability engineering 9
remediation 166, 168
repairs 13
    costs 54
    or replacement 28
replacement 23
    costs 28
resilient flooring 266
retention ponds 267
return on investment (ROI) 21, 62
reversing valve 129
Reynold's theory 83

Reynolds theory 100
ROI 26
rolling element bearings 93, 98, 99
roll roofing 107
roof 10, 75
 consultant 104, 270
 drains 75
 inspection 119
 penetrations 113
 warranties 116
roofing 42
 consultant 75
room preventive maintenance checklist 285
RP 214
RPM 87, 96
RPP 214
RPZ 214
RTUs 126
rubber flooring 266
rubber roofs 108
run to failure 15
rust 180, 181
 converter 181, 187
rust inhibitive paints 181

S

SAE weight 89
Safety, health, and environmental compliance 16
safety inspections 42
safety relief valve 216
salvage value 33
savings 21
Saybolt universal seconds 89
SBS (styrene butadiene styrene) 107
scale buildup 142
schedules 57
scope 48
sealed bearings 100
sealers 182
self lubricated 99
semi-hermetic compressors 131, 133
septic tanks 272
service life 41
service manuals 49
sewer gas 161
SHE 16
shellac 183
sick building syndrome 135, 166, 168
sidewalks 261
sight glass 132
silicones 176
single-ply membrane 107
skills 67, 68, 72
sleeve bearing 98
smoke and fire doors 229
smoke detector 273
 testing log 287
soap 92
sodium soap 93
solenoid valves 256
solid lubricants 92
solids 176
solvent 176
spalling 193
spherical roller bearings 96
split system 130
spreadsheet 58
squirrel cage fans 137
stack effect 166
stain killers 183
stains 185
standardizing 17
standards 39

standing seam metal roof 104, 109
storage 242
  of paints 190
storm water 267
strippers 266
stucco 184
stuffing box 99
suction 127
sump pumps 274
supplemental heat 130
surface prep 185
surfacing material 263
suspicion 169
swimming pools 274
symptoms 162, 163, 168, 170
synthetic oil 92
synthetic resins 176

T

tackiness 194
tamper signaling devices 243
tannic acid 187
tapered roller bearing 96
Teflon® 92, 93, 99
temperature 163, 170, 191
templates 55
tension 156
termite inspections 276
thermal imaging 9, 10, 122, 269
thermoplastics 104, 107
thermostatic expansion valve 128
third-party consultants 76
three Ps 159
time-based 9
titanium dioxide 177
tobacco smoke 160
tons 126
tools 73
topcoat 182
torch down 107
toxins 167
training 21, 69, 70, 71
transformers 230
tribology 83
troubleshooter's creed 288
trucks 280
truisms 289
trust but verify 74
tuberculosis 171
TXV 128

U

ultrasonic detection 9
under coat 182
unique identifier 44
urethane paints 180
useful life 33
  of oil 94
useful service life 38

V

vacuum cleaners 277
valves 278
vee-belts 149
vehicles 280
ventilation 164, 165, 166
vibration analysis 9
vinyl composition tile 266
viscosity 86, 87, 89
  index 90
volatile organic compounds (VOCs) 161, 188

W

walk-off mats 219, 266
walk-through 41, 42
warranty 51, 75
washers 254

watershed 268
water heaters 215
water softeners 281
wear factor 9
web browser 54
web page 55
wet cooling tower 139
wet insulation 121
wick 94
Willis Haviland Carrier 126
wood 184
  fiberboard 112
wool 99
work orders 15, 28, 48, 51, 53
wrinkling 194

Y

yellowing 194

Z

zeolite 281
zerk 96
  fitting 97
zinc 181
  oxide 181
  paints 181